Prions

Current Progress in Advanced Research

Edited by

Akikazu Sakudo

Laboratory of Biometabolic Chemistry
University of the Ryukyus
Okinawa
Japan

and

Takashi Onodera

Research Center for Food Safety
The University of Tokyo
Tokyo
Japan

Caister Academic Press

Caister Academic Press
Norfolk, UK

www.caister.com

British Library Cataloguing-in-Publication Data
A catalogue record for this book is available from the British Library

ISBN: 978-1-908230-24-9

Cover design adapted from Figure 5.1.

Printed and bound in Great Britain

Contents

Contributors

Hermann C. Altmeppen
Institute of Neuropathology
University Medical Center Hamburg-Eppendorf
Hamburg
Germany

h.altmeppen@uke.de

Yasuhisa Ano
Central Laboratories for Frontier Technology
Kirin Holdings Co. Ltd
Yokohama-shi, Kanagawa
Japan

yasuhisa_ano@kirin.co.jp

Frank Dohler
Institute of Neuropathology
University Medical Center Hamburg-Eppendorf
Hamburg
Germany

f.dohler@uke.de

Clemens Falker
Institute of Neuropathology
University Medical Center Hamburg-Eppendorf
Hamburg
Germany

c.falker@uke.de

Markus Glatzel
Institute of Neuropathology
University Medical Center Hamburg-Eppendorf
Hamburg
Germany

m.glatzel@uke.uni-hamburg.de

Richard Knight
The National CJD Research and Surveillance Unit
Western General Hospital
Edinburgh

r.knight@ed.ac.uk

Susanne Krasemann
Institute of Neuropathology
University Medical Center Hamburg-Eppendorf
Hamburg
Germany

skrasema@uke.de

Gerald McDonnell
STERIS limited
Basingstoke
UK

gerry_mcdonnell@steris.com

Shigeru Matsuda
Laboratory of Biometabolic Chemistry
School of Health Sciences
Faculty of Medicine
University of the Ryukyus
Okinawa
Japan

aKatsuKiway@yahoo.co.jp

Takashi Onodera
Graduate School of Agricultural and Life Sciences and Research Center for Food Safety
University of Tokyo
Tokyo
Japan

aonoder@mail.ecc.u-tokyo.ac.jp

Berta Puig
Institute of Neuropathology
University Medical Center Hamburg-Eppendorf
Hamburg
Germany

b.puig-martorell@uke.de

Suehiro Sakaguchi
Division of Molecular Neurobiology
Institute for Enzyme Research (KOSOKEN)
University of Tokushima
Tokushima
Japan

sakaguch@ier.tokushima-u.ac.jp

Akikazu Sakudo
Laboratory of Biometabolic Chemistry
School of Health Sciences
Faculty of Medicine
University of the Ryukyus
Okinawa
Japan

sakudo@med.u-ryukyu.ac.jp

Hideharu Shintani
School of Science and Engineering
Chuo Univertsity
Tokyo
Japan

shintani@mail.hinocatv.ne.jp

Katsuaki Sugiura
Graduate School of Agricultural and Life Sciences and Department of Global Animal Science
University of Tokyo
Tokyo
Japan

aksugiur@mail.ecc.u-tokyo.ac.jp

Keiji Uchiyama
Division of Molecular Neurobiology
Institute for Enzyme Research (KOSOKEN)
University of Tokushima
Tokushima
Japan

ku200@ier.tokushima-u.ac.jp

Takashi Yokoyama
Prion Disease Research Center
National Institute of Animal Health
Ibaraki
Japan

tyoko@affrc.go.jp

Preface

The discovery of prions (proteinaceous infectious particles) by Dr. Stanley Prusiner in 1982 initiated more than 30 years of intensive research into the biology of these agents. During this time significant knowledge of prion diseases, prion agents and prion protein (PrP) has accumulated as well as the development of major technical advances for detecting and analysing disease-related proteins.

The motivation behind this book is to provide students, scientists and engineers with recent progress of advanced research into prion biology. This book consists of 11 chapters; the first chapter (Chapter 1) is an overview of prions and prion diseases. The following four chapters (Chapters 2–5) deal with fundamental aspects of prion biology including PrP functions and molecular mechanisms of prion diseases. The next three chapters (Chapters 6–8) focus on clinical aspects of human prion diseases and current approaches for therapy and effective sterilization. The last part of the book summarizes animal prion diseases. In the final chapter, Professor Onodera discusses the likely future direction of research in this field.

It was an honor to have been given the opportunity to work with such eminent scientists as chapter contributors. Their efforts have made this book possible. In particular, I would like to acknowledge my co-editor, Professor Takashi Onodera, for stimulating discussion on prion biology during the compilation of this book. Finally, I wish to thank Sarah Collins and the other editorial staff at Caister Academic Press for their professionalism and dedication.

Akikazu Sakudo

Introduction

1

Takashi Onodera and Akikazu Sakudo

Abstract

Prion diseases or transmissible spongiform encephalopathies (TSEs) are fatal neurological diseases that include Creutzfeldt–Jakob disease (CJD) in humans, scrapie in sheep and goats, bovine spongiform encephalopathy (BSE) in cattle, and chronic wasting disease (CWD) in cervids. A key event in prion diseases is the conversion of the cellular, host-encoded prion protein (PrP^C) to its abnormal isoform (PrP^{Sc}) predominantly in the central nervous system of the infected host (Aguzzi *et al.*, 2004).The diseases are transmissible under some circumstances, but unlike other transmissible disorders, prion diseases can also be caused by mutations in the host gene. The mechanism of prion spread among sheep and goats that develop natural scrapie is unknown. CWD, transmissible mink encephalopathy (TME), BSE, feline spongiform encephalopathy (FSE), and exotic ungulate encephalopathy (EUE) are all thought to occur after the consumption of prion-infected material.

Most cases of human prion disease occur from unknown reasons, and > 20 mutation in the prion gene may lead to inherited prion disease. In other instances, prion diseases are contracted by exposure to prion infectivity. This raises the question of how a mere protein aggregate can trespass mucosal barriers, circumvent innate and adoptive immunity, and travel across the blood–brain barrier to eventually provoke brain disease. To start the chapters of this book we will introduce a few topics in current prion studies.

Risk assessment in food safety and blood products

Over the past years European Food Safety Authority (EFSA) has been requested several times to provide scientific advice and in particular to risk assessment in the area of animal TSEs [i.e. BSE, classical scrapie, atypical scrapie and chronic wasting disease (CWD) in ruminants], including specific questions on their zoonotic potential. In particular, EFSA's advice has been asked on some occasions by the European Commission in order to provide updated information on the zoonotic potential of TSEs other than BSE.

In the past, EFSA and the former Scientific Steering Committee (SSC) of the European Commission were asked to reflect on the zoonotic potential of CWD and possible role of cervids in the transmission of TSEs to humans. The SSC in 2003 concluded that a theoretical risk for prion transmission to humans consuming products of CWD affected cervids of all ages in countries where CWD exists cannot be excluded. The SSC also concluded that the early and widespread involvement of tissues in CWD infected animals did not allow defining SRM list, neither to define any lower age cut off has been defined for cattle in relation to BSE. Later, an EFSA opinion (EFSA, 2004) concluded that even though human TSE exposure risk through consumption of game from European cervids could be assumed to be minor, if at all existing, no final conclusion could be drawn due to the overall lack of scientific data.

Even in the field of vCJD, several developments have occurred recently [EFSA Panel on Biological Hazards (BIOHAZ), 2011], as written below.

- A case report described a 30-year-old man who died in January 2009 with symptoms suggestive of vCJD. This individual had a genotype (heterozygote case) previously not associated with disease. This created concern that there may be a second wave of vCJD cases in humans with a different genetic background.
- The identification of a further possible vCJD infection in the spleen of a patient with haemophilia A in UK raised the possibility that vCJD infection can be transmitted from person to person through the use of plasma-derived products.

Based on these developments, a number of questions were raised by European Commission: should current assumption on the number of people that may develop vCJD in the future be reviewed? How does this impact on the current assumptions regarding transmissibility through blood transfusion and tissue/cells transplantation? Does this change the number of individuals at risk of developing vCJD following a transfusion/transplantation? Are there measures to reduce any possible increased risk? Authors in this book will try to explain the current situations with their newest knowledge.

Prion clearance

Progressive accumulation of PrP^{Sc} can only occur if conversion of PrP^{C} to PrP^{Sc} is faster than PrP^{Sc} clearance. Therefore studying the clearance of prions is arguably as important as studying their generation. *Prnp*$^{-/-}$ mice develop more or less normally yet cannot replicate prions, making them a perfect model to study the half-life of the prion. Upon inoculation, residual infectivity all but disappears within 4 days, indicating that prions – commonly regarded as the sturdiest pathogens on earth – can be cleared *in vivo* with astonishing efficiency and speed. The identification of molecules and cells involved in prion clearance will be of great importance for therapeutics of prion diseases (Aguzzi *et al.*, 2012).

Neprilysin is a metalloprotease known to degrade extracellular amyloid such as Aβ. However, mice lacking or overexpressing neprilysin show no changes in prion pathogenesis. Therefore, prion clearance may be effected by extracellular proteases other than neprilysin, or by different mechanisms altogether (Glatzel *et al.*, 2005). In organotypic cerebellar slices, the pharmacogenetic ablation of microglia led to a 15-fold increase in prion titres (Falsig *et al.*, 2008), suggesting that microglias are the primary important effector of prion clearance. But how can microglia identify prions as edible materials? Milk fat globule epidermal growth factor 8 (Mfge8), a bridging molecule mediating phagocytosis of apoptotic cells, may represent a crucial link. Mfge8$^{-/-}$ mice showed accelerated prion pathogenesis, accompanied with reduced clearance of cerebral apoptotic bodies and increased PrP^{Sc} accumulation and prion titre (Kranich *et al.*, 2010), suggesting Mfge8-mediated prion clearance in prion-infected mouse brain. Interestingly, these were observed in C57BL/6x129Sv but not in C57BL/6 genetic background. Therefore, besides Mfge8, other molecules involved in phagocytosis of apoptotic cells could have potential to clear prions *in vivo*. These new molecules involved in the pathogenesis in prion diseases are worthwhile for putting more effort on them. These new molecules will be discussed further in a latter part of this book.

References

Aguzzi, A., and Polymenidou, M. (2004). Mammalian prion biology: one century of evolving concepts. Cell *116*, 313–327.

Aguzzi, A., and Zhu, C. (2012). Five questions on prion diseases. Plos Pahog. *8*, e1002651.

EFSA (European Food Safety Authority). (2004). Opinion of European Food Safety Authority on a surveillance programme for Chronic Wasting Disease in the European Union. EFSA J. *70*, 1–7.

EFSA Panel on Biological Hazards (BIOHAZ). (2011). Joint scientific opinion on any possible epidemiological or molecular association between TSEs in animals and humans. EFSA J. *9*, 1945–2055.

Falsig, J., Julius, C., Margalith, I., Schwarz, P., Heppner, F.I., and Aguzzi, A. (2008). A versatile prion replication assay in organotypic brain slices. Nat. Neurosci. *11*, 109–117.

Glatzel, M., Mohajeri, M.H., Poirier, R., Nitsch, R.M., Schwarz, P., Lu, B., and Aguzzi, A. (2005). No influence of amyloid-beta-degrading neprilysin activity on prion pathogenesis. J. Gen. Virol. *86*, 1861–1867.

Kranich, J., Krautler, N.J., Falsig, J., Ballmer, B., Li, S., Hutter, G., Schwartz, P., Moos, R., Julius, C., Miele, G., *et al.* (2010). Engulfment of cerebral apoptotic bodies controls the course of prion diseases in a mouse strain-dependent manner. J. Exp. Med. *207*, 2271–2281.

Prion Protein and the Family Members, Doppel and Shadoo

2

Akikazu Sakudo

Abstract

Prion diseases are devastating neurodegenerative disorders caused by infectious proteinaceous agents known as prions. Prion protein (PrP) gene (*Prnp*)-deficient mice do not infect with prion agent, indicating essential role of PrP for prion diseases. An abnormal isoform of prion protein (PrP), known as PrP^{Sc}, which is converted from cellular PrP (PrP^{C}), is thought to constitute the prion agent. Recently, proteins homologous to PrP have been found, suggesting the existence of other PrP family members, which so far include PrP, Doppel (Dpl) and Shadoo (Sho). In this chapter, the author introduces recent research on the physiological function of PrP and PrP-related proteins together with our own studies.

Prion protein

The prion hypothesis has been put forward to explain the pathogenesis of prion diseases (Prusiner, 1998). This hypothesis states that the prion protein (PrP) can exist in the normal cellular isoform (PrP^{C}), or in an 'infectious' abnormal isoform (PrP^{Sc}) that causes disease by converting the normal isoform into the abnormal isoform. There is significant conservation of sequence among mammalian PrPs. Moreover, PrP^{C} is highly expressed in neurons suggesting that the protein plays an important biological role. However, the physiological function of PrP^{C} remains unclear. Human PrP^{C} comprises a signal peptide (residues 1–22), an octapeptide repeat region (OR) (repeat of PHGGGWGQ, residues 51–91), a hydrophobic region (HR, residues 106–126), three peptide sequences forming an α-helix structure (H1-H3), and a glycosylphosphatidylinositol (GPI) anchor (residues 231–253) (Sakudo *et al.*, 2006). There are two consensus motifs for N-linked glycosylation (T181 and T197). In addition, a disulfide bond between Cys179 and Cys214 is essential for proper folding of PrP^{C}. Among these structural components, important regions for the physiological function of PrP are thought to be the GPI anchor, OR and HR. The GPI anchor is required for localizing the protein to cholesterol-rich lipid rafts within the cell membrane. Therefore, PrP^{C} is thought to be involved in signal transduction on lipid rafts as is the case with other GPI anchored proteins (Mouillet-Richard *et al.*, 2000). Interestingly, the GPI anchor plays an important role in the pathogenesis of prion diseases because transgenic mice expressing a mutant form of PrP in which the GPI anchor has been deleted do not show any symptoms after prion infection (Chesebro *et al.*, 2005).

During metabolic processing, PrP is cleaved between amino acid 110 and 111 within the HR region by ADAM (a disintegrin and metalloprotease), which is a family of peptidase proteins, such as ADAM10 or ADAM17 (Vincent *et al.*, 2001; Mangé *et al.*, 2004). In addition, cleavage at OR occurs during the generation of reactive oxygen species (ROS) (Watt *et al.*, 2005). The cleavage that occurs at the HR region is called alpha-cleavage, whereas cleavage at OR is called beta-cleavage (Mangé *et al.*, 2004; Watt *et al.*, 2005). Moreover, copper binds to OR at a histidine residue (Brown *et al.*, 1997) and induces endocytosis of PrP (Pauly *et al.*, 1998). Glycosaminoglycan also binds OR, but the significance of this binding event remains unclear (Gonzalez-Iglesias *et al.*, 2002). More interestingly, PrP having an extended OR causes inherited prion diseases (Kaski *et al.*, 2011; Jansen *et al.*, 2011).

Doppel and Shadoo

Doppel (Dpl), encoded by *Prnd*, and Shadoo (Sho), encoded by *Sprn*, are two new members of the PrP family of proteins (Fig. 2.1) (Watts *et al.*, 2007a). Given that the function of PrP remains elusive, information obtained from comparative studies of structural and functional analyses are of great interest. In addition, the relationship between this family of proteins and the pathogenesis of prion disease is a very important topic. However, Dpl is localized in the brain for only a limited duration in the developmental process. Moreover, the level of Dpl expression does not correlate with the onset of prion disease (Weissmann *et al.*, 1999; Tuzi *et al.*, 2002). These observations suggest there is no relationship between Dpl and prion diseases. By contrast, the expression of Sho decreases with the accumulation of PrP^{Sc} after prion infection (Watts *et al.*, 2007b). This decrease in the level of Sho, combined with a lowering in the level of PrP^{C} after prion infection, may make the cell particularly susceptible to oxidative stress. Nonetheless, PrP is the main target for the onset of prion diseases. Dpl shows the highest level of expression in the testis of wild-type mice. In addition, *Prnd*-knockout mice show sterility due to immature sperm (Behrens *et al.*, 2002). PrP is also expressed in testis, but unlike other cells the PrP is truncated at its C-terminus. The

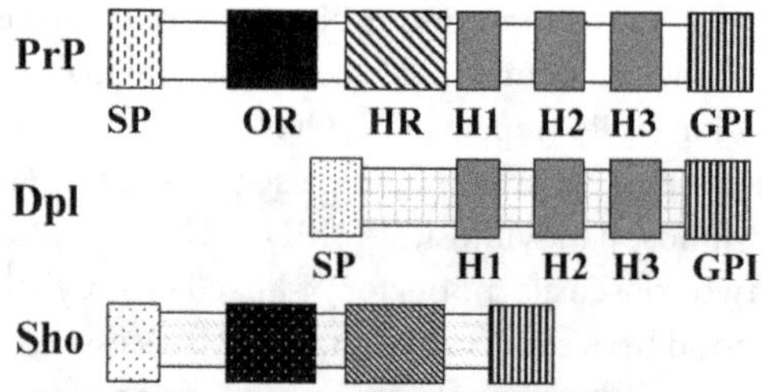

Figure 2.1 Structure of PrP family proteins. So far, PrP family proteins include PrP, Doppel (Dpl) and Shadoo (Sho). Dpl is similar to the C-terminal region of PrP, while Sho is similar to the N-terminal region of PrP. OR and HR play an essential role in the anti-apoptotic function of PrP. Sho also contains a tetra-repeat (TR) that displays similarity to OR. Indeed, Sho is thought to have an analogous function to that of PrP. Sho was recently identified by searching the database for homologous proteins to the HR of PrP. Although the signal peptide (SP) of Sho has not been found, the report that Sho is attached to cellular membranes via a GPI anchor suggests this region must exist. However, there is no helix structure or disulfide bond in Sho. H1–H3, helix regions. Modified from Sakudo *et al.* (2011) with permission from the Society for Antibacterial and Antifungal Agents, Japan.

N-terminal region of PrP resembles Dpl, suggesting that these proteins share a common function in testis (Shaked *et al.*, 1999). Sho protein levels were decreased in the brains of *Prnp*^{a} and *Prnp*^{b} mice (Westaway *et al.*, 2011), hamsters, voles and sheep infected with natural and experimental prion strains (Watts *et al.*, 2011). In addition, time course experiments showed that the levels of PrP^{Sc} versus Sho protein were inversely proportional (Watts *et al.*, 2011). Membrane anchoring and the N-terminal domain of PrP both influenced the inverse relationship between PrP^{Sc} and Sho (Watts *et al.*, 2011). By contrast, increased expression of Sho did not influence prion replication (Watts *et al.*, 2011), suggesting Sho merely acts as a marker for prion disease. Therefore, it remains unclear how Sho contributes to the pathogenesis of prion diseases. Regardless, depletion of Sho appears to be unimportant in terms of triggering prion diseases and in the processing and degradation of PrP^{Sc}. Recent studies using *Sprn*-regulated reporter mice has shown that *Sprn* is expressed in brain, thymus, heart, lung, liver, kidney, spleen, intestine, muscle and gonads in adult (Young *et al.*, 2011a). *Sprn* is also expressed during the developmental process in the embryo (Young *et al.*, 2011b). Interestingly, the expression profile of *Sprn* in testis and ovary resemble that of *Prnp*. Moreover, *in vitro* analysis demonstrated that Sho has neuroprotective activity and suppressive activity against the toxicity of Dpl and N-terminal region deleted PrP (Watts *et al.*, 2007). Interestingly, Sho is highly expressed in the central nervous system (CNS), similar to PrP, and has a similar structure to the N-terminal part of PrP (Watts *et al.*, 2007a). The corresponding region of PrP (i.e. OR) in Sho includes a tetra repeat (TR) motif that is rich in Arg and Gly residues. Although OR is well known as a copper-binding region, it is unclear whether copper or other metals bind TR.

Our research using PrP gene-deficient neuronal cells

So far, there are six types of *Prnp*-deficient mice; ZrchI, Npu, ZrchII, Ngsk, Rikn and Rcm0 (Sakudo *et al.*, 2006). However information obtained from *Prnp*-knockout mice did not clearly reveal PrP function due to the generation of obscure/subtle phenotypes or poor reproducibility. Thus, our group tried to apply a simpler system to PrP analysis (Sakudo *et al.*, 2003, 2006).

Firstly, we established a *Prnp*-deficient neuronal cell line HpL derived from the hippocampal region of *Prnp*-knockout mouse embryo using a retrovirus vector including the oncogene SV40 large T antigen gene. Similarly, a HW neuronal cell line derived from wild-type mice was established. Interestingly, HW extended neurites and the cells differentiated after serum deprivation. By contrast, HpL showed apoptosis, which was suppressed by reintroduction of *Prnp* (Fig. 2.2). As serum deprivation is one of the indicators of oxidative stress, we compared the levels of intracellular oxidative stress in PrP-expressing and non-expressing cells in a time-dependent manner. Generation of oxidative stress induced by serum deprivation was suppressed by PrP expression (Sakudo *et al.*, 2003). Moreover, activity of superoxide dismutase (SOD), which eliminates the superoxide anion, was increased in *Prnp*-reintroduced cells (Sakudo *et al.*, 2003). These results suggest that PrP enhances anti-oxidative defence activity and prevent cell death induced by oxidative stress.

PrP contains two highly conserved regions; OR and HR. We generated a series of deletion mutants to elucidate the important part of the molecule responsible for the anti-apoptotic function of PrP (Fig. 2.3) (Sakudo *et al.*, 2005a). PrPΔ#1, in which the OR region was deleted (residues 53–94), and PrPΔ#2, in which the N-terminal half of

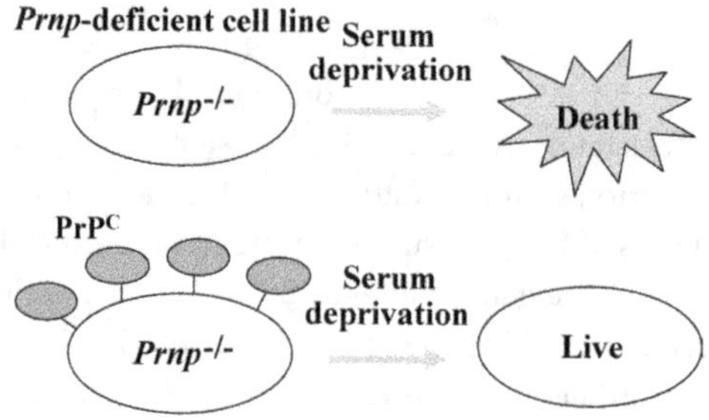

Figure 2.2 PrP gene-deficient cell line (serum deprivation). PrP gene (*Prnp*)-deficient neuronal cells (*Prnp*$^{-/-}$) deprived of serum undergo apoptotic cell death, which is suppressed by reintroduction of *Prnp* (encoding PrP). Modified from Sakudo *et al.* (2011) with permission from the Society for Antibacterial and Antifungal Agents, Japan.

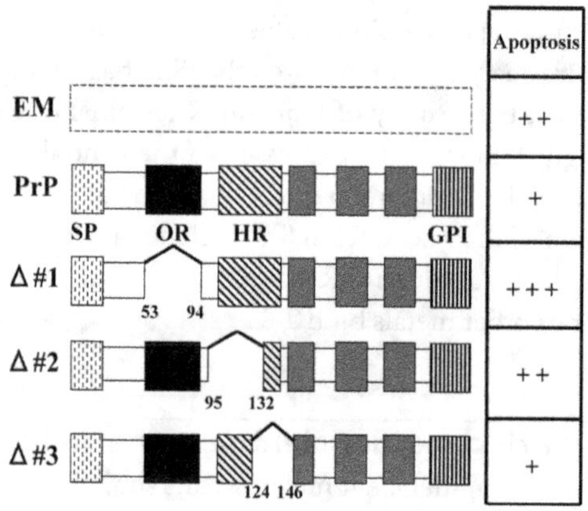

Figure 2.3 Deletion of OR or N-terminal half of HR decreases anti-apoptotic activity of PrP. PrP gene-deficient neuronal cells transfected with empty vector (EM) undergo cell death when deprived of serum. However, the same cell line transfected with PrP-expressing vector show decreased levels of apoptosis compared to cells transfected with EM. By contrast, cell death cannot be prevented when the same experiment is conducted with deletion mutants of PrP, such as Δ#1 (OR-deleted PrP) and Δ#2 (N-terminal half of HR deleted PrP). The Δ#3 (C-terminal half of HR deleted PrP) retains anti-apoptotic activity. Modified from Sakudo *et al.* (2011) with permission from the Society for Antibacterial and Antifungal Agents, Japan.

the HR region was deleted (residues 95–132), both lost anti-apoptotic activity. However, the anti-apoptotic activity was retained by PrPΔ#3, in which the C-terminal half of HR (residues 124–146) was deleted. Therefore, both OR and the N-terminal half of HR are essential for the anti-apoptotic activity of PrP. Furthermore, because OR binds copper, PrP also regulates copper metabolism. In addition, HR binds stress inducible protein 1 (STI1). The binding of PrP to both copper and STI1 are important for its SOD activation (Sakudo *et al.*, 2005b). Taken together, these results suggest PrP prevents neuronal cell death and appears to be indispensable for sustaining cell survival. The OR and HR regions are responsible for binding to copper and STI1, which triggers the activation of SOD and thereby prevents cell death.

Conclusion

It is interesting to analyse the relationship among PrP family proteins (i.e. PrP, Dpl and Sho) in terms of their pathogenicity and physiological function. Recently, there have been *Sprn*-knockout and *Prnp*/*Sprn*-double knockout mice available. The studies have shown that the knockout and double knockout mice survived to over 600 days of age without prion infection, while lack of *Sprn* did not affect prion incubation time (Daude *et al.*, 2012). Nonetheless, *Sprn*-knockout, *Prnp*/*Sprn*-double knockout, and *Prnp*/*Prnd*/*Sprn*-triple knockout mice would provide invaluable information on the function of these proteins.

In the brains of prion-infected animals there is a deficiency in PrP^C and an accumulation of PrP^{Sc}, which is induced by the conversion of PrP^C into PrP^{Sc}. Thus far, PrP^C is believed to play a role in anti-oxidative stress and appears to be involved in the mechanism by which the accumulation of PrP^{Sc} increases oxidative stress. Conversely, a deficiency of PrP^C decreases anti-oxidative stress. It will be interesting to establish whether Dpl and Sho are also involved in oxidative stress metabolism. Although Dpl and Sho expression do not modulate prion diseases, further analysis is required to determine whether Dpl and Sho indirectly contribute to prion diseases via regulation of oxidative stress or other systems. Currently, all prion diseases are fatal with no effective form of treatment. However, information related to the pathogenesis of prion diseases may provide information that could be used in the development of a novel form of therapy.

Acknowledgement

A part of this review was published in a recent Japanese review (Sakudo *et al.*, 2011). The Society for Antibacterial and Antifungal Agents, Japan kindly gave us permission to use figures from the published review, some of which are presented in this study in a modified form (Sakudo *et al.*, 2011). This work was supported by Grant-in-Aids for Scientific Research from the Ministry of Education, Science, Culture and Technology of Japan (23780299), Grants-in-Aid from the Research Committee of Prion disease and Slow Virus Infection, the Ministry of Health, Labour and Welfare of Japan, and Grant-in-Aid for Government and Administration related research from Senri Life Science Foundation.

References

Behrens, A., Genoud, N., Naumann, H., Rülicke, T., Janett, F., Heppner, F.L., Ledermann, B., and Aguzzi, A. (2002). Absence of the prion protein homologue Doppel causes male sterility. EMBO J. *21*, 3652–3658.

Brown, D.R., Qin, K., Herms, J.W., Madlung, A., Manson, J., Strome, R., Fraser, P.E., Kruck, T., von Bohlen, A., Schulz-Schaeffer, W., *et al.* (1997). The cellular prion protein binds copper *in vivo*. Nature *390*, 684–687.

Chesebro, B., Trifilo, M., Race, R., Meade-White, K., Teng, C., LaCasse, R., Raymond, L., Favara, C., Baron, G., Priola, S., *et al.* (2005). Anchorless prion protein results in infectious amyloid disease without clinical scrapie. Science *308*, 1435–1439.

Daude, N., Wohlgemuth, S., Brown, R., Pitstick, R., Gapeshina, H., Yang, J., Carlson, G.A., and Westaway, D. (2012). Knockout of the prion protein (PrP)-like Sprn gene does not produce embryonic lethality in combination with PrP(C)-deficiency. Proc. Natl. Acad. Sci. U.S.A. *109*, 9035–9040.

González-Iglesias, R., Pajares, M.A., Ocal, C., Espinosa, J.C., Oesch, B., and Gasset, M. (2002). Prion protein interaction with glycosaminoglycan occurs with the formation of oligomeric complexes stabilized by Cu(II) bridges. J. Mol. Biol. *319*, 527–540.

Mangé, A., Béranger, F., Peoc'h, K., Onodera, T., Frobert, Y., and Lehmann, S. (2004). Alpha- and beta-cleavages of the amino-terminus of the cellular prion protein. Biol. Cell. *96*, 125–132.

Mouillet-Richard, S., Ermonval, M., Chebassier, C., Laplanche, J.L., Lehmann, S., Launay, J.M., and Kellermann, O. (2000). Signal transduction through prion protein. Science *289*, 1925–1928.

Jansen, C., Voet, W., Head, M.W., Parchi, P., Yull, H., Verrips, A., Wesseling, P., Meulstee, J., Baas, F., van Gool, W.A., *et al.* (2011). A novel seven-octapeptide repeat insertion in the prion protein gene (*PRNP*) in a Dutch pedigree with Gerstmann-Sträussler-Scheinker disease phenotype: comparison with similar cases from the literature. Acta Neuropathol. *121(1)*, 59–68.

Kaski, D.N., Pennington, C., Beck, J., Poulter, M., Uphill, J., Bishop, M.T., Linehan, J.M., O'Malley, C., Wadsworth, J.D., Joiner, S., *et al.* (2011). Inherited prion disease with 4-octapeptide repeat insertion: disease requires the interaction of multiple genetic risk factors. Brain. *134(Pt 6)*, 1829–1838.

Pauly, P.C., and Harris, D.A. (1998). Copper stimulates endocytosis of the prion protein. J. Biol. Chem. *273*, 33107–33110.

Prusiner, S.B. (1998). Prions. Proc. Natl. Acad. Sci. U.S.A. *95*, 13363–13383.

Sakudo, A., and Tanaka, Y. (2011). Prion protein (PrP) and the family members. Bokin-Bobai-Shi *39*, 77–81.

Sakudo, A., Lee, D.C., Saeki, K., Nakamura, Y., Inoue, K., Matsumoto, Y., Itohara, S., and Onodera, T. (2003). Impairment of superoxide dismutase activation by N-terminally truncated prion protein (PrP) in PrP-deficient neuronal cell line. Biochem. Biophys. Res. Commun. *308*, 660–667.

Sakudo, A., Lee, D.C., Nishimura, T., Li, S., Tsuji, S., Nakamura, T., Matsumoto, Y., Saeki, K., Itohara, S., Ikuta, K., *et al.* (2005a). Octapeptide repeat region and N-terminal half of hydrophobic region of prion protein (PrP) mediate PrP-dependent activation of superoxide dismutase. Biochem. Biophys. Res. Commun. *326*, 600–606.

Sakudo, A., Lee, D.C., Li, S., Nakamura, T., Matsumoto, Y., Saeki, K., Itohara, S., Ikuta, K., and Onodera, T. (2005b). PrP cooperates with STI1 to regulate SOD activity in PrP-deficient neuronal cell line. Biochem. Biophys. Res. Commun. *328*, 14–19.

Sakudo, A., Onodera, T., Suganuma, Y., Kobayashi, T., Saeki, K., and Ikuta, K. (2006). Recent advances in clarifying prion protein functions using knockout mice and derived cell lines. Mini Rev. Med. Chem. *6*, 589–601.

Shaked, Y., Rosenmann, H., Talmor, G., and Gabizon, R.A. (1999). C-terminal-truncated PrP isoform is present in mature sperm. J. Biol. Chem. *274*, 32153–32158.

Tuzi, N.L., Gall, E., Melton, D., and Manson, J.C. (2002). Expression of doppel in the CNS of mice does not modulate transmissible spongiform encephalopathy disease. J. Gen. Virol. *83*, 705–711.

Vincent, B., Paitel, E., Saftig, P., Frobert, Y., Hartmann, D., De Strooper, B., Grassi, J., Lopez-Perez, E., and Checler, F. (2001). The disintegrins ADAM10 and TACE contribute to the constitutive and phorbol ester-regulated normal cleavage of the cellular prion protein. J. Biol. Chem. *276*, 37743–37746.

Watt, N.T., Taylor, D.R., Gillott, A., Thomas, D.A., Perera, W.S., and Hooper, N.M. (2005). Reactive oxygen species-mediated beta-cleavage of the prion protein in the cellular response to oxidative stress. J. Biol. Chem. *280*, 35914–35921.

Watts, J.C., and Westaway, D. (2007a). The prion protein family: diversity, rivalry, and dysfunction. Biochim. Biophys. Acta. *1772*, 654–672.

Watts, J.C., Drisaldi, B., Ng, V., Yang, J., Strome, B., Horne, P., Sy, M.S., Yoong, L., Young, R., Mastrangelo, P., *et al.* (2007b). The CNS glycoprotein Shadoo has PrP(C)-like protective properties and displays reduced levels in prion infections. EMBO J. *26*, 4038–4050.

Watts, J.C., Stöhr, J., Bhardwaj, S., Wille, H., Oehler, A., Dearmond, S.J., Giles, K., and Prusiner, S.B. (2011). Protease-resistant prions selectively decrease Shadoo protein. PLoS Pathog. *7*, e1002382.

Weissmann, C., and Aguzzi, A. (1999). Perspectives: neurobiology. PrP's double causes trouble. Science *286*, 914–915.

Westaway, D., Genovesi, S., Daude, N., Brown, R., Lau, A., Lee, I., Mays, C.E., Coomaraswamy, J., Canine, B., Pitstick, R., *et al.* (2011). Down-regulation of Shadoo in prion infections traces a pre-clinical event inversely related to PrP(Sc) accumulation. PLoS Pathog. *7*, e1002391.

Young, R., Le Guillou, S., Tilly, G., Passet, B., Vilotte, M., Castille, J., Beringue, V., Le Provost, F., Laude, H., and Vilotte, J.L. (2011a). Generation of Sprn-regulated reporter mice reveals gonadic spatial expression of the prion-like protein Shadoo in mice. Biochem. Biophys. Res. Commun. *412*, 752–756.

Young, R., Bouet, S., Polyte, J., Le Guillou, S., Passet, B., Vilotte, M., Castille, J., Beringue, V., Le Provost, F., Laude, H., *et al.* (2011b). Expression of the prion-like protein Shadoo in the developing,mouse embryo. Biochem. Biophys. Res. Commun. *416*, 184–187.

Function of Cellular Prion Protein

3

Takashi Onodera, Katsuaki Sugiura, Shigeru Matsuda and Akikazu Sakudo

Abstract

Although much is known about the effect of PrP^{Sc} in prion diseases, the normal function of PrP^{C} is poorly understood. PrP^{C} may act as an antiapoptotic agent by blocking some of the internal environmental factors that initiate apoptosis. PrP-knockout methods provide powerful hints on the neuroprotective function of PrP^{C}. Using PrP^{C}-knockout cell lines, the inhibition of apoptosis through STI1 is mediated by PrP^{C}-dependent SOD activation. Recently several reports show that PrP^{C} participate in trans-membrane signalling process associated with haematopoietic stem cell replication and neuronal differentiation. Besides PrP-knockout exhibited wide spread alterations of oscillatory activity in the olfactory bulb as well as altered paired-pulse plasticity at the dendrodendritic synapse. Both the behavioural and electro-physiological phenotypes could be rescued by neuronal PrP^{C} expression.

Neuprotein Shadoo (Sho), similarly to PrP^{C}, can prevent neuronal cell death induced by the expression of PrP△HD mutants, an artificial PrP mutant devoid of internal hydrophobic domain. Sho can efficiently protect cells against ecitotoxin-induced cell death by glutamates. Sho and PrP seem to be dependent on similar domains, in particular N-terminal (N) and their internal hydrophobic domain. ShoΔN and ShoΔHD displayed a reduced stress-protective activity but are complex glycosylated and attached to outer leaflet of the plasma membrane via GPI anchor indicating that impaired activity is not due to incorrect cellular trafficking. In Shadoo overexpressed mice showed large amyloid plaques not seen in wild-type mice. However Shadoo is not a major modulator of PrP^{Sc} accumulation and scrapie pathogenesis. Sho and PrP share a stress-protective activity. The ability to adopt a toxic conformation of PrP^{Sc} seems to be specific for PrP.

PrP^{C} protects neurons from stress-induced apoptosis

Neurogenesis

Recently, several reports showed that PrP^{C} participate in trans-membrane signalling process associated with haematopoietic stem cell replication and neuronal differentiation (Mouillet-Richard *et al.*, 2000; Zhang *et al.*, 2006; Steele *et al.*, 2006). Abundant expression of PrP^{C} has been detected during mouse embryogenesis in association with the developing nervous system (Manson *et al.*, 1992; Miele *et al.*, 2003; Tremblay *et al.*, 2007). In the developing mouse brain, undifferentiated neural progenitor cells in the mitotically active ventricular zone do not express PrP^{C}. In contrast, post-mitotic neurons express high levels of PrP^{C} after their last mitosis in the neuroepithelium as migrate towards marginal layers and differentiate (Steele *et al.*, 2006; Tremblay *et al.*, 2007). Thus, PrP^C may be expressed exclusively in differentiated neurons. Studies *in vitro* have showed that expression of PrP^C is positively correlated with differentiation of multipotent neuronal precursors into mature neurons (Steele *et al.*, 2006). In addition, treatment of embryonic hippocampal neurons with recombinant PrP^C enhance neurite outgrowth and survival (Kanaani *et al.*, 2005).

The distribution of PrP^C in the developing nervous system of cattle (Peralta *et al.*, 2011), as well as in mice (Tremblay *et al.*, 2007) and humans (Adle-Biassette *et al.*, 2006) suggest that PrP^C plays a functional role in neural development. While PrP-null mice displays no overt in neural phenotype (Beuler *et al.*, 1992), numerous subtle phenotypes have been reported (Steele *et al.*, 2007), including reduction in the number of neural precursor cells in developing mouse embryo (Steele *et al.*, 2006). Other studies have shown that PrP^C induced neuritogenesis in embryonic hippocampal neurons cultured *in vitro* (Kanaani *et al.*, 2005; Lopes *et al.*, 2005). PrP^C interacts with stress-inducible protein 1 (STI1) (Zanata *et al.*, 2002), which is a heat shock protein (Lassle *et al.*, 1997). The interaction of PrP^C with STI1 not only activate cyclic adenosine monophosphate (cAMP)-dependent protein kinase A to transducer a survival signal but also induces phosphorylation/activation of the mitogen-activated protein kinase to promote neuritogenesis (Lopes *et al.*, 2005). The expression of mammalian PrP^C in the neuroepithelium and its spatial and temporal relation with neural marker nestin and MAP-2 also suggest the participation of PrP^C in the process of neural differentiation during early embryogenesis (Peralta *et al.*, 2011). The use of ES cells to study the potential role or PrP^C will indicate how PrP^C is up-regulated during the differentiation of stem/progenitor cells.

Neuroprotection

The mammalian cellular prion protein (PrP^C) is a highly conserved glycoprotein localized in membrane lipid rafts and anchored to cell surface by glycophosphatidylinositol (GPI) (McKinley *et al.*, 1991). It is present in many cell types, and is particularly abundant in neurons (Taraboulos *et al.*, 1992). Under certain conditions PrP^C may undergo conversion into a conformationally altered isoform (scrapie prion protein or PrP^{Sc}) widely believed to be the pathogenic agent in prion disease or transmissible spongiform encephalopathies (TSE) (Caughey *et al.*, 1991; Pan *et al.*, 1993). Although much is known about the effect of PrP^{Sc} in prion diseases, the normal function of PrP^C is poorly understood. PrP^C has alpha and beta-cleavage site during normal processing and host translational modifications (Mange *et al.*, 2004). The most commonly observed function on PrP^C is copper-binding. The octapeptide-repeat region of PrP^C binds with Cu^{2+} within the physiological concentration range (Hornshaw *et al.*, 1995; Kramer *et al.*, 2001; Prusiner 1997; Miura *et al.*, 1999; Zeng *et al.*, 2003). Furthermore, PrP^C display a functional role in normal brain metabolism of copper (Brown *et al.*, 1997). Besides binding with Cu^{2+} at the synapse, PrP^C serves as Cu^{2+} buffer as well (Kretzschmar *et al.*, 2000). Overexpression of PrP^C increases Cu^{2+} uptake into cells (Brown, 1999), while PrP^C-knockout mouse show a lower synaptosomal Cu^{2+} concentration than normal mice (Kretzschmar *et al.*, 2000). On the other hand, the Cu^{2+} rapidly and reversibly stimulate the internalization of PrP^C during PrP^C endocytosis (Pauly *et al.*, 1998; Kubosaki *et al.*, 2003; Haigh *et al.*, 2005). Through the binding with Cu^{2+}, PrP^C displays superoxide dismutase (SOD) activity *in vitro* (Brown *et al.*, 1999; Vasallo *et al.*, 2003). Interestingly, treatment with copper chelator cuprizone induce TSE-like spongiform degeneration (Pattison *et al.*, 1973). Therefore, Cu^{2+} metabolism appears to play an important role in not only PrP function but also the pathogenesis of prion diseases.

PrP^C may act as an antiapoptotic agent by blocking some of the internal or environmental factors that initiate apoptosis (Bounhar *et al.*, 2001; Roucou *et al.*, 2005). Mature PrP^C trend to localize in lipid raft of cells (Taraboulos *et al.*, 1992). As lipid rafts are membrane structure specialized in signalling, a potential role of PrP^C in signal transduction may be anticipated. Discovery of several PrP^C-interacting candidates has facilitated understanding of PrP^C function (Table 3.1). PrP^C-interacting molecules are most likely involved in signal transduction. In addition, a phosphorylating function of PrP^C, mediated by caveolin-1 to indirectly increase Fyn (a member of Src family of tyrosine kinase) phosphorylation, governs the downstream production of NADPH oxidase-dependent reactive oxygen species and activation of the extracellular regulated kinase 1/2 has been demonstrated (Mouillet-Richard *et al.*, 2000; Schneider *et al.*, 2003). PrP^C interact with normal phosphoprotein synapsin Ib and cytoplasmic adaptor protein Grb2 without being deciphered with prion interactor Pint1 (Spielhaupter *et al.*, 2001). Bovine PrP strongly interact with the catalytic α/α' subunit of protein kinase CK2 to increase the phosphotransferase activity of CK2, thus leading to the phosphorylation of calmodulin (Maggio *et al.*, 2000).

Recently, PrP^C has been demonstrated to modulate serotonergic receptor-signalling in the inducible serotonergic 1C115-HT cell line, namely modulation of 5-hydroxytryptamine (5-HT) receptor coupling to activate G-protein functions, as well as acting as a protagonist to promote homeostasis of serotonergic neurons (Moulliet-Richard *et al.*, 2005). In addition, PrP^C binds with extracellular matrix laminin to promote genesis and maintenance of neurites (Graner *et al.*, 2000a,b). In fact, a recent study has discovered PrP^C to induce self-renewal of long term populating haematopoietic stem cells (Zhang *et al.*, 2006). Furthermore, another study has revealed that PrP is expressed on the multipotent neural precursors and mature neurons without being detected in glia, suggesting that PrP^C plays an important role in neural differentiation (Steele *et al.*, 2006). Therefore, the interaction between PrP^C and various signal transduction molecules speaks well for important (such as differentiation and cell survival) in the living system.

PrP-knockout methods provide useful hints on the neuroprotective function of PrP^C (Sakudo *et al.*, 2006). A *Prnp*-deficient cell line (HpL3–4), immortalized from hippocampal neuronal precursors, is sensitive to serum deprivation-induced apoptosis but is activated/survived with PrP^C expression (Kuwahara *et al.*, 1999). Overexpression of Bcl-2 in this cell line reveals a functional relation of PrP^C with Bcl-2 in the anti-apoptotic pathway (Kurschner *et al.*, 1995; Kuwahara *et al.*, 1999). Prevention of cell death in cultured retinal explants from neonatal rats and mice induced by anisomycin (a protein synthesis inhibitor) unfurls and effect associated with PrP^C–STI1 interactions (Zanata *et al.*, 2002). The production of another type of heat-shock protein (Hsp 70) is enhanced when PrP levels elevate during hyperglycaemia (Shyu *et al.*, 2005). According to finding in another study, the inhibition of apoptosis through STI1 is mediated by PrP^C-dependent SOD activation (Sakudo *et al.*, 2005). The functional role of STI1 and PrP^C have been confirmed in both murine and bovine systems (Hashimoto *et al.*, 2000). The late onset of severe ataxia and loss of cerebellar Purkinje cells in several knockout mouse lines (Sakaguchi *et al.*, 1996; Moore *et al.*, 1999; Rossi *et al.*, 2001) suggest a lack of protection of cerebellum by PrP^C in these mice. Interestingly, deposition of PrP^{Sc} has been located in the deep cerebellar nuclei (DCN) of scrapie-infected sheep (Ersdal *et al.*, 2003). Future studies with a microarray analysis (Park

Table 3.1 Proteins interacting with PrP

Proteins	Methods	References
Stress-inducible protein1	Complementary hydropathy	Martins *et al.* (1997)
Tubulin	Cross-linking by bis(sulfosuccinimidyl)-suberate	Nieznanski *et al.* (2005)
Neural adhesion molecule (N-CAM)	Cross-linking by formaldehyde	Schmitt-Ulms *et al.* (2001)
Dystroglycan	Detergent-dependent immunoprecipitation	Keshet *et al.* (2000)
Neuronal isoform of nitric oxide synthase (nNOS)	Detergent-dependent immunoprecipitation	Keshet *et al.* (2000)
Grp94	Immunoprecipitation	Capellari *et al.* (1999)
Protein disulphide isomerase	Immunoprecipitation	Capellari *et al.* (1999)
Calnexin	Immunoprecipitation	Capellari *et al.* (1999)
Calreticulin	Immunoprecipitation	Capellari *et al.* (1999)
ZAP-70	Immunoprecipitation	Mattei *et al.* (2004)
NF-E2-related factor 2 (Nrf2)	Interaction with PrP23–231-alkaline phosphatase probe	Yehiely *et al.* (1997)
Amyloid precursor protein-like protein 1 (Aplp1)	Interaction with PrP23–231-alkaline phosphatase probe	Yehiely *et al.* (1997)
F-box protein-6	Interaction with PrP23–231-alkaline phosphatase probe	Yehiely *et al.* (1997)
Neural F-box protein 42kDa (NFB42)	Interaction with PrP23–231-alkaline phosphatase probe	Yehiely *et al.* (1997)
Postsynaptic density 95kDa (PSD-95)/SAP-90 associated protein	Interaction with PrP23–231-alkaline phosphatase probe	Yehiely *et al.* (1997)
Protein tyrosine phosphatase, non-receptor type-21	Interaction with PrP23–231-alkaline phosphatase probe	Yehiely *et al.* (1997)
Predicted protein KIAA0443	Interaction with PrP23–231-alkaline phosphatase probe	Yehiely *et al.* (1997)
Glial fibrillary acidic protein (GFAP)	Interaction with radioisotope-labelled PrP27–30	Oesch *et al.* (1990)
Hsp60 of *Brucella abortus*	Pull-down assay	Watarai *et al.* (2003)
Bcl-2	Yeast two-hybrid system	Kurschner and Morgan (1995)
Heat shock protein 60kDa	Yeast two-hybrid system	Edenhofer *et al.* (1996)
37kDa Laminin receptor protein (LRP)	Yeast two-hybrid system	Rieger *et al.* (1997)
Pint1	Yeast two-hybrid system+ immunoprecitation	Spielhaupter *et al.* (2001)
SynapsinIb	Yeast two-hybrid system+ immunoprecitation	Spielhaupter *et al.* (2001)
Neuronal phosphoprotein Grb2	Yeast two-hybrid system+ immunoprecitation	Spielhaupter *et al.* (2001)
Neurotrophin receptor interacting MAGE homologue	Yeast two-hybrid system+*in vitro* binding assay+immunoprecipitation	Bragason *et al.* (2005)
Potassium channel tetramerization domain containing one (KCTD1) protein	Yeast two-hybrid system	Huang *et al.* (2012)

Proteins	Methods	References
Rab7a	Coimmunoprecipitation+ immunofluorescence	Zafar *et al.* (2011)
Rab9	Coimmunoprecipitation+ immunofluorescence	Zafar *et al.* (2011)
HS-1-associated protein X-1 (HAX-1)	Yeast two-hybrid system	Jing *et al.* (2011)
Histone H1	Far Western immunoblotting	Strom *et al.* (2011)
Histone H3	Far Western immunoblotting	Strom *et al.* (2011)
Lamin B1	Far-western immunoblotting	Strom *et al.* (2011)
14-3-3beta protein	Immunoprecipitation+pull-down assays	Liu *et al.* (2010)
Casein kinase II	Immunoprecipitation+pull-down assays	Chen *et al.* (2008)
Tetraspanin-7	Yeast two-hybrid system+ immunoprecipitation	Guo *et al.* (2008)
2P domain K+channel TREK-1 protein	Bacterial two-hybrid+ immunoprecipitation	Azzalin *et al.* (2006)
ADAM23	Immunoprecipitation+pull-down assay	Costa *et al.* (2009)

et al., 2006) applied in eye-blink conditioning of mice may provide insight to understanding the normal function of PrP^C in the DCN of cerebellum.

A loss of PrP^C function could be implicated in the pathogenesis of prion diseases and PrP^C-dependent pathways might be involved in neurotoxic signalling (Table 3.1). For example, *in vivo* crosslinking of PrP^C by antibodies triggered neuronal apoptosis (Solforosi *et al.*, 2004) and PrP^C-dependent receptors were postulated to explain neurotoxic effect of a PrP mutant lacking the hydrophobic domain (see next sections) (Winklhofer *et al.*, 2008).

Taken together, PrP^C is functionally involved in copper metabolism, signal transduction, neuroprotection and cell maturation. Despite these putative roles, mice null for PrP^C display no consistent phenotype apart from complete resistance to TSE infection (Bueler *et al.*, 1992, 1993). Further search for PrP^C-interaction molecules using *Prnp*$^{-/-}$ mice and various types of *Prnp*$^{-/-}$ cell lines under various conditions may elucidate the PrP^C functions.

Synaptic plasticity

In $PrP^{-/-}$ mice Kim *et al.* (2007) have observed pathological alterations and some physiological dysfunctions in olfactory bulb (OB). Recently, Le Pichen *et al.* (2009) have uncovered a significant phenotype of $PrP^{-/-}$ mice in the olfactory system utilizing a combination of genetic, behavioural and physiological and physiological techniques in a systems approach. They employed so-called 'cookie finding task', a test of broad olfactory acuity, to analyse a battery of mice including PrP knockout on multiple genetic backgrounds and transgenic mice in which *Prnp* expression was driven by cell type-specific promoters. $PrP^{-/-}$ mice exhibited impaired behaviour that was rescued in transgenic mice expressing PrP^C specifically in neurons but not in mice expressing only extraneuronal PrP^C. $PrP^{-/-}$ mice displayed altered behaviour in an additional olfactory test (habituation–dishabituation) which was also rescued by transgenic neuronal PrP expression suggesting that the phenotype was olfactory specific.

Besides, the odour-evoked electrophysiological properties of the OB of PrP knockouts were studied (Le Pichon *et al.*, 2009). In these mice, alterations in the patterns of oscillatory activity in the OB were detected. The plasticity of dendrodendritic synaptic transmission was altered between granule cells and mitral cell. Le Pichon *et al.* propose that electrophysiological alterations at the dendrodendritic synapse in the OB could underlie the behaviour phenotypes.

In detail, the cookie-finding phenotype was manifest in three $PrP^{-/-}$ lines (Zurich I PrP knockout: Beuler *et al.*, 1992; Nagasaki PrP knockout: Sakaguchi *et al.*, 1996; Edinburgh PrP knockout: Manson *et al.*, 1994) on alternate genetic backgrounds, strong evidence of its dependence on PrP^C rather than other genetic factors. PrP knockouts also displayed altered behaviour in the habituation–dishabituation task, suggesting the phenotype was likely olfactory-specific. $PrP^{-/-}$ mice exhibited wide spread alterations of oscillatory activity in the OB as well as altered paired-pulse plasticity at the dendrodendritic synapse. Importantly, both the behavioural and electrophysiological phenotypes could be rescued by neuronal PrP^C expression. These data suggest a critical role for PrP^C in the normal processing of sensory information by the olfactory system.

Disruption was observed in local field potential (LFP) oscillation and in the plasticity of the dendrodendritic synapse, either, or both, of which could contribute to the $PrP^{-/-}$ behavioural phenotype. Oscillatory LFPs may act to organize information flow within the olfactory system (Lledo *et al.*, 2006; Stopher *et al.*, 2007) by constraining the timing of mitral cell action potentials (Kasiwadani *et al.*, 1999). In addition, gamma oscillations are specifically implicated in behavioural performance in olfactory tasks (Nusser *et al.*, 2001; Brown *et al.*, 2005; Beshel *et al.*, 2007). Therefore, alterations in oscillatory timing during odour exposure may perturb OB output to higher centres by disrupting how information is packaged within a breathing cycle.

Altering the dendrodendritic synapse may have multiple functional consequence. This synapse may mediate lateral inhibition between ensembles of mitral cells, and be critical for olfactory discrimination (Yokoi *et al.*, 1995; Urban 2002). Additionally, because granule cells receive convergent information onto their proximal dendritic arbour from multiple higher brain areas (Shepherd, 2003), disruption of the dendrodendritic synapse may alter the transmission of centrifugal modulation of OB mitral cells.

High-frequency oscillations in the OB (gamma and high-gamma) are shown *in vitro* to result from the rapid and reciprocal interactions between granule and mitral cells across the dendrodendritic synapse (Schoppa *et al.*, 2006; Lagier *et al.*, 2007). Therefore, Le Picheon's observation could imply that increased facilitation of mitral cell inhibitory postsynaptic potential (IPSP) following repetitive spiking decreases the dynamic range and increases the duration of gamma oscillations across the boundaries of breath. Although both oscillatory and synaptic effects could be reversed by neuronal PrP^C expression, they cannot claim a causal link between these findings.

Mitral cell receive facilitated inhibition in $PrP^{-/-}$ mice. This facilitation could result from either pre- and/or post-synaptic changes to the dendrodendritic synapse. Future work should determine the precise synaptic localization of the PrP^C protein as well as its biochemical interactions with synaptic machinery. Furthermore, the transgenic rescue strategy cannot indicate whether the observed phenotypes result from the developmental changes in olfactory circuitry. Future use of conditional strategies using tissue specific promoters may allow a more precise dissection of the physiological and behavioural importance of PrP^C for olfactory processing.

Myelination

A late-onset peripheral neuropathy has been identified in PrP^C-deficient Nagasaki (*Prnp*$^{Ngsk/Ngsk}$) and Zurich-I (*Prnp*$^{0/0}$) mice (Bueler *et al.*, 1992; Sakaguchi *et al.*, 1996; Nishida *et al.*, 1999). This indicates that PrP^C might have a role in peripheral neuropathies. At 60 weeks of age, all investigated *Prnp*$^{0/0}$ mice ($n = 52$) showed chronic demyelinating polyneuropathy (CDP) (Bremer *et al.*, 2010). CDP was 100% penetrant and conspicuous in all investigated peripheral nerves (sciatic and trigeminal nerves, dorsal and ventral spinal roots). Besides, CDP was associated with another two independently targeted *Prnp* knockout mouse lines, *Prnp*$^{GFP/GFP}$ (Heikenwalder *et al.*, 2008) mice and *Prnp*$^{Edbg/Edbg}$ (Manson *et al.*, 1994) mice.

Prnp$^{0/0}$ and *Prnp*$^{Edbg/Edbg}$ mice suffered from CDP despite normal expression of Doppel (Dpl) (Moore *et al.*, 1999), indicating that Dpl regulation did not cause polyneuropathy. CDP was present in mice lacking both *Prnp* and *Prnd* (the gene for Dpl) (Genoud *et al.*, 2004), but absent from mice selectively lacking *Prnd* (Behrens *et al.*, 2002). Therefore Dpl is not required for the maintenance of peripheral nerves. PrP^C might interact with myelin component directly or through other axonal proteins. Some of the reported PrP^C interacting proteins have roles in homeostasis (Rutinshauser *et al.*, 2009), and represent possible candidates for mediation of its myelinotrophic effects. The octapeptide repeat region was not required for myelin maintenance, whereas mice PrP lacking central domain (aa 94–134) developed CDP (F. Baumann, 2007). The hydrophobic core, but not the charge cluster (CC_2), of this central PrP^C domain was essential for peripheral myelin maintenance.

PrP^C undergoes regulated proteolysis in late secretory compartments (McMahon *et al.*, 2001; Watt *et al.*, 2005; Sunyach *et al.*, 2007; Walmsley *et al.*, 2009). Bremer *et al.* (2010) 0bserved an association between the presence of CDP and lack of C1 fragment in sciatic nerves. All PrP mutants which CDP was rescued produced abundant C1. Cleavage of PrP^C appeared therefore to be linked to its myelinotrophic function. This conjuncture might also explain the requirement for membrane anchoring of PrP^C uncovered in ^{tg}GPI$^-$PrP mice (Chesebro *et al.*, 2005), as anchorless PrP^C did not undergo regulated proteolysis.

Prion diseases mainly affect the CNS, myelin degeneration in optic nerves, corpus callosum or spinal cords was not detected in 60-week-old *Prnp*$^{0/0}$ mice (Bremer *et al.*, 2010). Nevertheless subliminal myelin pathologies might extend to central myelin in *Prnp*$^{0/0}$ mice (Nazor *et al.*, 2007), and transgenic mice expressing toxic PrP^C show both peripheral and central myelinopathy (Radovanovic *et al.*, 2005; Baumann *et al.*, 2007). PrP^C deficiency was reported to affect synaptic function (Collinge *et al.*, 1994; Mallucci *et al.*, 2002). However, the amplitudes of foot muscle compound action potentials following distal stimulation were not significantly altered in 53-week-old *Prnp*$^{0/0}$ mice arguing against an important synaptic defect in neuromuscular synaptic junction.

It has been suggested that PrP^C has various roles in immunity (Issacs *et al.*, 2006), and lymphocytes are important in mouse models of hereditary demyelinating neuropathies. As the CDP in our mutant mice was not modulated by removal of *Rag1*, lymphocytes are not involved in its pathogenesis. The combined results of restricting expression of PrP^C of neurons and of selectively depleting PrP^C from neurons indicate that the expression of PrP^C by the neuron is essential for the long-term integrity of peripheral myelin sheaths (Bremer *et al.*,

2010). Not only was the trophic function of PrP^C exerted *in trans*, but also correlated with the proteolytic processing of in diverse transgenic mouse models. These findings identify PrP^C as a critical messenger of transcellular axomyelinic communication and indicate that regulated proteolysis of axonal PrP^C might exposed domains that interact with Schwann cell receptors. Clarifying the molecular basis of these phenomena might lead better understanding of peripheral neuropathies – particularly those of late onset – and might help to uncover new therapeutic targets.

PrP^C mediates toxic signalling by PrP^{Sc}

Mice with prion disease show misfolded PrP accumulation and develop extensive neurodegeneration, in contrast to mouse models of Alzheimer's disease (AD) or Parkinson's disease (PD), in which neuronal loss is rare. Therefore prion infected mice allow access to mechanism linking protein misfolding to neuronal death. Mallicci's group have previously shown rescue of neuronal loss and reversal of early cognitive and morphological changes in prion-infected mice by depleting PrP in neurons, preventing prion replication and abrogating neurotoxicity (Mallucci *et al.*, 2003; Mallucci *et al.*, 2007; White *et al.*, 2008). Recently, the same group have shown that PrP^{Sc} replication causes sustained unfolded protein response (UPR) induction with persistent, deleterious expression of eLF2α-P in prion disease (Moreno *et al.*, 2012). The resulting chronic blockade of protein synthesis leads to synaptic failure, spongiosis and neuronal loss. Promoting eLF2-P dephosphorylation rescues vital translation rates and is thereby neuroprotective, whereas preventing this further reduces translation and enhances neurotoxicity. The data support the development of generic proteostatic approaches to therapy in prion (Balch *et al.*, 2008; Tsaytler *et al.*, 2011). The unfolded PrP^C response works as protective cellular mechanism triggered by rising levels of misfolded PrP^{Sc} protein (Moreno *et al.*, 2012).

In another study, expression of PrP^C in neuronal cells is required to mediate neurotoxic effects of PrP^{Sc} (Chesebro *et al.*, 2005). PrP^{Sc} might elicit a deadly signal through a PrP^C dependent signalling pathway. Spontaneous neurodegeneration in transgenic mice expressing a PrP mutant without the N-terminal ER-targeting sequence indicated a toxic potential of PrP when located in cytosolic compartment (cytoPrP) (Ma *et al.*, 2002). Toxicity of cytoPrP seems to be dependent on its association with cellular membranes (Wang *et al.*, 2006) and its binding to Bcl-2, an antiapoptotic protein present at the cytosolic side of ER and mitochondrial membranes (Rambold *et al*, 2006). Might the toxic potential of misfolded PrP in the cytosol be relevant to the pathogenesis of prion diseases? Most recent information revealed an impairment of the ubiquitin-proteasome system (UPS) in prion-infected mice. In conjunction with *in vitro* and cell culture approaches, it was proposed that prion neurotoxicity is linked to PrP^{Sc} oligomers, which translocate to the cytosol and inhibit the URS (Kristiansen *et al.*, 2007).

Stress-inducible and toxic signalling mediated by PrP^C are interconnected

PrP^C expression is indispensable for prion-induced neurotoxicity (Brandner *et al.*, 1996), implying PrP^C could be a receptor for prions to trigger detrimental signalling. Strittmatter reported that PrP^C transduces the synaptic cytotoxicity of amyloid-β (A) oligomers *in vitro*

(Lauren *et al.*, 2009) and in Aβ transgenic mice (Gimbel *et al.*, 2010). Moreover, different anti-PrP antibodies or their antigen-binding fragment that disrupt the PrP–Aβ interaction were able to block the Aβ-mediated disruption of synaptic plasticity. These findings were important because they suggest the involvement of PrP^C in AD pathogenesis. However, others found that the absence of PrP^C did not prevent deficits in hippocampal-dependent behavioural tests upon intracerebral Aβ injection (Balducci *et al.*, 2010). It has been suggested that variations in copper availability may contribute these discrepancies (Stys *et al.*, 2012).

Parkin *et al.* (2007) reported an interaction between PrP^C and the rate-limiting enzyme in the production of Aβ, the β-secretase BACE1 and, more recently, two subsequent studies have also found direct links: PrP^C has been reported to be a receptor for Aβ oligomers (Lauren *et al.*, 2009) and the expression of PrP^C is controlled by the amyloid intracellular domain (AICD) (Vincent *et al.*, 2009). There are two potential roles suggested for PrP^C in AD: One, a role in the physiological regulation of amyloid precursor protein (APP) via interaction with BACE1; and two, a role in the pathological progression of AD by mediating Aβ toxicity by binding $Aβ_{42}$-oligomers. The feedback loop between, PrP^C, BACE1, APP and AICD are described, and provides a model linking these recent observations (Kellett *et al.*, 2009). However, several questions remains to be answered, including 'What effect does A $_{42}$-oligomer binding have on the functions of PrP^C?', 'How do the levels of PrP^C vary in the brains of AD patients and age-matched controls?' and 'What is the effect of altering PrP^C levels in mouse models of AD?'. Understanding the molecular and cellular mechanisms involved in the interactions between PrP^C and APP/Aβ is crucial to the understanding of AD pathogenesis.

PrP^C seems to regulate the β-secretase cleavage of amyloid precursor protein, thereby regulating the production of Aβ (Parkin *et al.*, 2007). In addition, α-secretase regulates the cleavage of PrP^C, regulating an N-terminal fragment with neuroprotective activity (Cisse *et al.*, 2005; Guillot-Sestier, *et al.*, 2009). PrP^C also binds to transmembrane proteins such as the 67-kDa laminin receptor (Rieger *et al.*, 1997; Gauczynski *et al.*, 2001; Hundt *et al.*, 2001), neural cell adhesion molecules (Schmitt-Ulms *et al.*, 2001; Santuccione *et al.*, 2005), G protein-coupled serotonergic receptors (Mouillet-Richard *et al.*, 2005), and low density lipoprotein receptor-related protein 1 (Taylor *et al.*, 2007; Parkyn *et al.*, 2008), which are able to promote intracellular signalling-mediated neuronal adhesion and differentiation as well as PrP^C internalization. Remarkably, PrP^C functions as receptor or co-receptor for extracellular matrix proteins such as laminin (Graner *et al.*, 2000a,b) and vitronectin (Hajj *et al.*, 2007), as well as STI1 (Zanata *et al.*, 2002). These data suggest that glycosylphosphatidylinositol-anchored PrP^C is a potential scaffold receptor in a multiprotein, cell surface, signalling complex, that may be the basis for the multiple neuronal functions ascribed to PrP^C (Linden *et al.*, 2008, 2009; Martins *et al.*, 2010).

In hippocampal neurons STI1-PrP^C engagement induces an increase in intracellular Ca^{2+} levels. Using a best candidate approach to test potential channels involved in Ca^{2+} influx, Beraldo *et al.* (2010) found that α-bungarotoxin, a specific inhibitor for α7 nicotinic acetylcholine receptor (α7nAChR), was able to block PrP^C-STI1-mediated signalling, neuroprotection, and neuritogenesis. STI1 can interact with the PrP^C·α7nAChR complex to promote signalling and provide potential target for modulation of the effect of prion protein in neurodegenerative diseases. The drugs that prevent bindings of Aβ 1–42 to α7nAChR seems to be beneficial in a model of AD (Wang *et al.*, 2009). It seems that STI1 binding to

PrP^C can hijack one of the key signalling pathways related to AD and learning and memory. Therefore, it is possible that STI1 modulation containing a complex containing PrP^C and α7nAChR may play an important role in AD.

Shadoo, a highly conserved glycoprotein with similarities to PrP^C

In the search for homologous/paralogues of PrP^C, a new gene was identified termed *Sprn*, encoding for a protein denoted Shadoo (Sho) (Premzl *et al.*, 2003). Sho is highly conserved from fish to mammals. The sequence homology between Sho and PrP is restricted to the internal hydrophobic domain; however, certain features, such as a N-terminal repeat region and a C-terminal GPI anchor, are conserved, suggesting that Sho and PrP may be functionally related. Experimental evidence for the post-translational modifications and cell surface localization of Sho was first presented for zebrafish Sho (Miesbauer *et al.*, 2006) and afterward also for mouse Sho (Watts *et al.*, 2007). Moreover, it was demonstrated that Sho, similarly to PrP^C, can prevent neuronal cell death induced by the expression of PrP△HD (hydrophobic domain) mutants, an artificial PrP mutant devoid of internal hydrophobic domain (Watts *et al.*, 2007). Stress-protective activity of Sho is not restricted to counteracting the toxic effects of PrPΔHD. Sakthivelu *et al.* (2011) employed glutamate as a physiologically relevant stressor to show Sho can efficiently protect cells against excitotoxin-induced cell death. Depletion mutants revealed that the stress-protective activity of Sho and PrP seems to be dependent on similar domains, in particular the N-terminal and their internal hydrophobic domain. ShoΔN (N-terminal) and ShoΔHD displayed a reduced stress- protective activity but are complex glycosylated and attached to the outer leaflet of the plasma membrane via GPI anchor, indicating that the impaired activity is not due to incorrect cellular trafficking.

The N-terminal domain of PrP is intrinsically disordered, and it has been shown that intrinsically disordered domains are involved in protein–protein interactions (Tompa *et al.*, 2009). Thus, it will be an attractive idea to assume that the N-terminal domains of PrP^C and Sho mediate interaction with a yet unknown co-receptor required for intracellular signal transmission. A circular dichroism analysis of recombinant Sho indicated that the whole protein might be unstructured (Watts *et al.*, 2007). The HD is the only domain with significant sequence homologies between Sho and PrP^C. The HD prompted dimerization of both Sho and PrP^C and was part of dimer interface. It is worth mentioning that dimerization is a common features of many cell surface receptors; therefore, it can be speculated that dimer formation is involved in signal transmission of PrP^C and Sho- dependent pathways. Formations of both PrP and Sho homodimers is not dependent on *N*-linked glycosylation and already occurs during transit through the secretory pathway. Similarly, PrP and Sho mutants devoid of the N-terminal domains are still capable of forming homodimers. However, they have an impaired stress-protective activity.

Sho is stress-protective, but does not mediate PrP^{Sc}-induced toxicity

Expression of murine *Sprn* transgene significantly increased brain Sho protein levels in all three mouse generated (Wang *et al.*, 2011). Following infection with mouse-adapted scrapie strain 22L, all transgenic mice tested exhibited characteristics of scrapie disease.

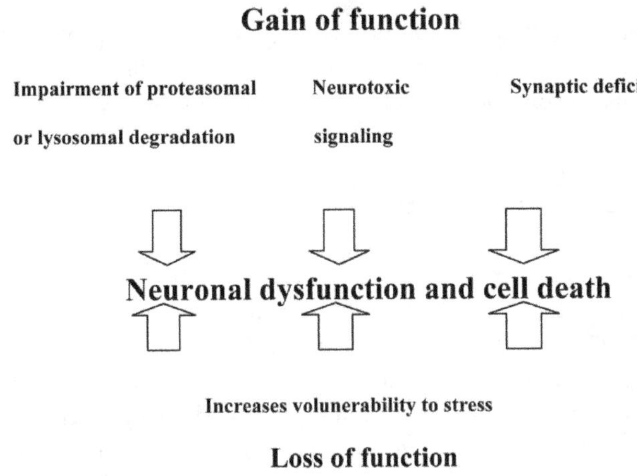

Figure 3.1 Gain and loss of function in prion disease.

Importantly, there was no correlation between the expression level or incubation time of Sho with disease phenotypes. Although the function of Sho are as yet little characterized, the gain of function experiments seems to be essential for CNS development in mice. Wang *et al.* (2011) generated mice overexpressing Sho to determine the role of Sho in the pathogenesis of transmissible spongiform encephalopathy (TSE). Wang reported that Sho overexpression has no correlation with incubation period of scrapie disease or with disease progression. There is no possible relationship between quantitative levels of Sho expression and scrapie pathology.

To evaluate the survival time, 22L strain of scrapie was injected intracerebrally into the brains of wild-type and *Sprn* overexpressed mice with mouse PrP-promoter (TgMoSprn). All 16 prion-infected wild-type mice showed abnormal behaviour including tremors and ataxia by 85 days and all had died by 149 days. The disease incubation period in infected wild-type mice were not significantly different from these of infected TgMo*Sprn* mice; three lines totalled to 40 mice.

In Sho overexpressed transgenic mice, Wang *et al.* (2011) detected large amyloid plaques not seen in wild-type mice. Recent work has shown that reduction in levels of Sho was not a direct or simple consequence of PrP^{Sc} accumulation. Instead, Sho protein levels are specific for the inoculated TSE agent and were not an intrinsic and invariant host process (Miyazawa *et al.*, 2010). Overexpression of Sho does not affect PrP, indicating that Sho has an alternate function. Other studies have shown that Sho exhibit no clear protective role in infected mice (Jeffrey *et al.*, 1997; Lloyd *et al.*, 2009), although infection with MRL scrapie agent induced widespread spongiform lesions and corresponding reduction in the levels of Sho (Miyazawa *et al.*, 2010) with no reduction in the time from incubation to neurological disease (Gossner *et al.*, 2009). In PrP knockout mouse brain there was no significant change in expression of Sho (Watts *et al.*, 2007), further demonstrating that Sho protein and PrP protein are independent. The unaltered survival time of scrapie infected TgMo*Sprn* mice is not in accordance with a neuroprotective effect of Sho, but it is not completely ruled out as there might be possible interference with a Sho overexpressing phenotype. The altered lesion profiles in terminally sick TgMo*Sprn* mice, also not correlated neuroprotective function of Sho, although the analysis of semiquantitative lesion profile might be too insensitive to reveal minimal changes when used in a loss of function model restricted to a limited number of neurons. Anyway, Sho is not a major modulator of PrP^{Sc} accumulation and scrapie pathogenesis.

Sho mutants devoid of the internal hydrophobic domain do not acquire a toxic potential

Studies in transgenic mice revealed the unexpected finding that PrP can acquire a neurotoxic potential by deleting the internal hydrophobic domain (Shmerling *et al.*, 1998; Baumann *et al.*, 2007; Li *et al.*, 2007). The neurotoxic potential of PrPΔHD is independent of the propagation of infectious prions, a phenomenon also seen for other neurotoxic PrP mutants (Winklhofer *et al.*, 2008). Although the underlying mechanism of PrPΔHD -induced toxicity are still elusive, co-expression of wild-type PrP^C completely prevents toxic effects of PrPΔHD. Based on this intriguing observation, it has been hypothesized that stress-protective signalling of PrP^C and the neurotoxic signalling of PrPΔHD are transmitted through a common co-receptor, which remains to be identified (Shmerling *et al.*, 1998; Baumann *et al.*, 2007; Li *et al.*, 2007; Rambold *et al.*, 2008). Co-transfection experiment with PrP-deficient cerebellar granule neurons indicated that Sho has a PrP^C-like activity to alleviate toxic effects of PrPΔHD expression (Watts *et al.*, 2007). Sakthivelu *et al.* (2011) have been able to recapitulate the toxic activity of PrPΔHD expression in their cell culture model and demonstrate the protective activity of PrP and Sho against PrPΔHD-induced toxicity. In addition, Sakthivelu *et al.* (2011) showed that ShoΔHD lost its ability to protect against stress-induced cell death. However, ShoΔHD did not acquire a toxic activity, at least not under the experimental conditions tested. In summary, Sho and PrP share a stress-protective activity. However, the ability to adopt a toxic conformation seems to be specific for PrP.

References

Adle-Biassette, H., Verney, C., Peoc'h, K., Dauge, M.C., Razavi, F., Choudat, L., Gressens, P., Budka, H., and Hein, D. (2006). Immunohistochemical expression prion protein (PrP^C) in the human forebrain during development. J. Neuropathol. Exp. Neurol. *65*, 698–706.

Azzalin, A., Ferrara, V., Arias, A., Cerri, S., Avella, D., Pisu, M.B., Nano, R., Bernocchi, G., Ferretti, L., and Comincini, S. (2006). Interaction between the cellular prion (PrP^C) and the 2P domain K+ channel TREK-1 protein. Biochem. Biophys. Res. Commun. *346*, 108–115.

Balch, W.E., Morimoto, R.I., Dillin, A., and Kelly, J.W. (2008). Adapting proteostasis for disease intervention. Science *319*, 916–919.

Balducci, C., Beeg, M., Stravalaci, M., Bastone, A., Sclip, A., Biasini, E., Tapella, L., Colombo, L., Manzoni, C., Borsello, T., *et al.* (2010). Synthetic amyloid-beta oligomers impair long-term memory independently of cellular prion protein. Proc. Natl. Acad. Sci. U.S.A. *107*, 2295–2300.

Baumann, F., Tolnay, M., Brabeck, C., Pahnke, J., Kloz, U., Niemann, H.H., Heikenwalder, M., Rulicke, T., Burkle, A., and Aguzzi, A. (2007). Lethal recessive myelin toxicity of prion protein lacking its central domain. EMBO J. *26*, 538–547.

Behrens, A., Genoud, N., Naumann, H., Rulicke, T., Janett, F., Heppner, F.L., Ledermann, B., and Aguzzi, A. (2002). Absence of the prion protein homologue Doppel causes male sterility. EMBO J. *21*, 3652–3658.

Beraldo, F.H., Arantes, C.P., Santos, T.G., Queiroz, N.G., Young, K., Rylett, R.J., Markus, R.P., Prado, M.A., and Martins, V.R. (2010). Role of α7 nicotinic acetylcholine receptor in calcium signaling induced by prion protein interaction with Stress-inducible Protein 1. J. Biol. Chem. *285*, 36542–36550.

Beshel, J., Kopell, N., and Kay, L.M. (2007). Olfactory bulb gamma oscillations are enhanced with task demands. J. Neurosci. *27*, 8358–8365.

Bounhar, Y., Zhang, Y., Goodyer, C.G., and LeBlanc, A. (2001). Prion protein protects human neurons against Bax-mediated apoptosis. J. Biol. Chem. *276*, 39145–39149.

Bragason, B.T., and Palsdottir, A. (2005). Interaction of PrP with NRAGE, a protein involved in neuronal apoptosis. Mol. Cell Neurosci. *29*, 232–244.

Brandner, S., Isenmann, S., Raeber, A., Fischer, M., Sailer, A., Kobayashi, Y., Marino, S., and Weissmann, C. (1996). Normal host prion protein is necessary for scrapie-induced neurotoxicity. Nature *379*, 339–343.

Bremer, J., Baumann, F., Tiberi, C., Wessig, C., Fischer, H., Schwarz, P., Steele, A.D., Toyka, K.V., Nave, K.-A., Weis, J., *et al.* (2010). Axonal prion protein is required for peripheral myelin maintenance. Nat. Neurosci. *13*, 310–318.

Brown, D.R. (1999). Prion protein expression aids cellular uptake and veratridine-induced release of copper. J. Neurosci. Res. *58*, 717–725.

Brown, D.R., Qin, K., Herms, J.W., Madlung, A., Manson, J., Strome, R., Fraser, P.E., Kruck, T., Von Bohlen, A., Schulz-Schaeffer, W., *et al.* (1997). The cellular prion protein binds copper *in vitro*. Nature *390*, 684–687.

Brown, S.L., Joseph, J., and Stopfer, M. (2005). Encoding a temporally structured stimulus with a temporally structured neural representation. Nat. Neurosci. *8*, 1568–1576.

Bueler, H., Fischer, M., Lang, Y., Fluethmann, H., Lipp, H.P., DeArmond, S.J., Prusiner, S.B., Aguet, M., and Weissmann, C. (1992). Normal development and behavior of mice lacking the neuronal cell-surface PrP protein. Nature *356*, 577–582.

Bueler, H., Aguzzi, A., Sailer, A., Greiner, R.A., Autenried, P., Aguet, M., and Weissmann, C. (1993). Mice devoid of PrP are resistant to scrapie. Cell 73, 1339–1347.

Capellari, S., Zaidi, S.I., Urig, C.B., Perry, G., Smith, M.A., and Petersen, R.B. (1999). Prion protein glycosylation is sensitive to redox change. JBC *274*, 34846–34850.

Caughey, B.W., Dong, A., Bhat, K.S., Ernst, D., Hayes, S.F., and Caughey, W.S. (1991). Secondary structure analysis of the scrapie-associated protein PrP 27–30 in water by infrared spectroscopy. Biochemistry *30*, 7672–7680.

Chen, J., Gao, C., Shi, Q., Wang, G., Lei, Y., Shan, B., Zhang, B., Dong, C., Shi, S., Wang, X., *et al.* (2008). Casein kinase II interacts with prion protein *in vitro* and forms complex with native prion protein *in vivo*. Acta Biochim Biophys Sin (Shanghai). *40*, 1039–1047.

Chesebro, B., Trifilo, M., Race, R., Meadr-White, K., Teng, C., LaCasse, R., Raymond, L., Favara, C., Baron, G., Priola, S., *et al.* (2005). Anchorless prion protein results in infectious amyloid disease without clinical scrapie. Science *308*, 1435–1439.

Cisse, M.A., Sunyach, C., Lefranc-Jullien, S., Postina, R., Vincent, B., and Checler, F. (2005). The disintegrin ADAM9 indirectly contributes to the physiological processing of cellular prion by modulating ADAM10 activity. J. Biol. Chem. *280*, 40624–40631.

Collinge, J., Whittington, M.A., Sidle, K.C., Smith, C.J., Palmer, M.S., Clarke, A.R., and Jefferys, J.G. (1994). Prion protein is necessary for normal synaptic function. Nature *370*, 295–297.

Costa, M.D., Paludo, K.S., Klassen, G., Lopes, M.H., Mercadante, A.F., Martins, V.R., Camargo, A.A., Nakao, L.S., and Zanata, S.M. (2009). Characterization of a specific interaction between ADAM23 and cellular prion protein. Neurosci. Lett. *461*, 16–20.

Edenhofer, F., Rieger, R., Famulok, M., Wendler, W., Weiss, S., and Winnacker, E.L. (1996). Prion protein PrPc interacts with molecular chaperones of the Hsp60 family. J. Virol. *70*, 4724–4728.

Ersdal, C., Ulvund, M.J., Benestad, S.L., and Tianulis, M.A. (2003). Accumulation of pathogenic prion protein (PrPSc) in nervous and lymphoid tissue of sheep with subclinical scrapie. Vet. Pathol. *40*, 164–174.

Gauczynski, S., Peyrin, J.M., Haik, S., Leucht, C., Hundt, C., Rieger, R., Krasemann, S., Deslys, J.P., Dormont, D., Lasmezas, C.I., *et al.* (2001). The 37-kDa/67-kDa laminin receptor acts as the cell-surface receptor for the cellular prion protein. EMBO J. *20*, 5863–5875.

Genoud, N., Behrens, A., Miele, G., Robay, D., Heppner, F.L., Freigang, S., and Aguzzi, A. (2004). Disruption of Doppel prevents neurodegeneration in mice with extensive Prnp deletions. Proc. Natl. Acad. Sci. U.S.A. *101*, 4198–4203.

Gimbel, D.A., Nygaard, H.B., Coffey, E.E., Gunther, E.G., Lauren, J., Gimbel, Z.A., and Strittmatter, S.M. (2010). Memory impairment in transgenic Alzheimer mice require cellular prion protein. J. Neurosci. *30*, 6367–6374.

Gossner, A.G., Bennet, N., Hunter, N., and Hopkins, J. (2009). Differential expression of Prnp and Sprn in scrapie infected sheep also reveals Prnp genotype specific differences. Biochem. Biophys. Res. Commun. *378*, 862–866.

Graner, E., Mercadante, A.F., Zanata, S.M., Forlenza, O.V., Cabral. A.L., Veiga, S.S., Juliano, M.A., Roesler, R., Walz, R., Minetti, A., *et al.* (2000a). Cellular prion protein binds laminin and mediates neuritogenesis. Brain Res. Mol. Brain Res. *76*, 85–92.

Graner, E., Mercadante, A.F., Zanata, S.M., Martins, V.R., Jay, D.G., and Brentani, R.R. (2000b). Laminin-induced PC-12 cell differentiation is inhibited following laser inactivation of cellular prion protein. FEBS Lett. *482*, 257–260.

Guillot-Sestier, M.V., Sunyach, C., Druon, C., Scarzello, S., and Checler, F. (2009). The alpha-secretase-derived N-terminal product of cellular prion, N1, displays neuroprotective function *in vitro* and *in vivo*. J. Biol. Chem. *284*, 35973–35986.

Guo, M., Huang, T., Cui, Y., Pan, B., Shen, A., Sun, Y., Yi, Y., Wang, Y., Xiao, G., and Sun, G. (2008). PrPC interacts with tetraspanin-7 through bovine PrP154–182 containing alpha-helix 1. Biochem. Biophys. Res. Commun. *365*, 154–157.

Haigh, C.L., Edwards, K., and Brown, D.R. (2005). Copper binding is governing determinant of prion protein turnover. Mol. Cell Neurosci. *30*, 186–196.

Hashimoto, A., Onodera, T., Ikeda, H., and Kitani, H. (2000). Isolation and characterization of fetal bovine brain cells in primary culture. Res. Vet. Sci. *69*, 39–46.

Hajj, G.N., Lopes, M.H., Mercadante, A.F., Veiga, S.S., da Silveira, R.B., Santos, T.G., Ribeiro, K.C., Juriano, M.A., Jacchieri, S.G., Zanata, S.M., *et al.* (2007). Cellular prion protein interaction with vitronectin supports axonal growth and is compensated by integrins. J. Cell Sci. *120*, 1915–1926.

Heikenwalder, M., Kurrer, M.O., Margalith, I., Kranich, J., Zeller, N., Haybaeck, J., Polymenidou, M., Matter, M., Bremer, J., Jackson, W.S., *et al.* (2008). Lymphotoxin-dependent prion replication in inflammatory stromal cells of granulomas. Immunity *29*, 998–1008.

Hornshaw, A.F., McDermott, J.R., and Candy, J.M. (1995). Copper binding to the N-terminal tandem repeat regions of mammalian and avian prion protein. Boichem. Biophys. Res. Commun. *207*, 621–629.

Huang, T., Xu, J., Xiang, J., Lu, Y., Chen, R., Huang, L., Xiao, G., and Sun, G. (2012). PrPC interacts with potassium channel tetramerization domain containing 1 (KCTD1) protein through the PrP(51–136) region containing octapeptide repeats. Biochem. Biophys. Res. Commun. *417*, 182–186.

Hundt, C., Peyrin, J.M., Haik, S., Gauczynski, S., Leucht, C., Rieger, R., Riley, M.L., Deslys, J.P., Dormont, D., Lasmezas, C.I., *et al.* (2001). Identification of interaction domains of the prion protein with its 37-kDa/67-kDa laminin receptor. EMBO J. *20*, 5876–5886.

Issacs, J.D., Jackson, G.S., and Altmann, D.M. (2006). The role of the cellular prion protein in the immune system. Clin. Exp. Immunol. *146*, 1–8.

Jeffrey, M., Goodsir, C.M., Bruce, M.E., McBride, P.A., and Fraser, J.R. (1997). *In vivo* toxicity of prion protein in murine scrapie: ultrastructural and immunogold studies. Neuropathol. Appl. Neurobiol. *23*, 93–101.

Jing, Y.Y., Li, X.L., Shi, Q., Wang, Z.Y., Guo, Y., Pan, M.M., Tian, C., Zhu, S.Y., Chen, C., Gong, H.S., *et al.* (2011). A novel PrP partner HS-1 associated protein X-1 (HAX-1) protected the cultured cells against the challenge of H_2O_2. J. Mol. Neurosci. *45*, 216–228.

Kanaani, J., Prusiner, S.B., Diacovo, J., Baekkeskov, S., and Legname, G. (2005). Recombinant prion protein induces rapid polarization and development of synapses in embryonic rat hippocampal neurons *in vitro*. J. Neurochem. *95*, 1373–1386.

Kashiwadani, H., Sasaki, Y.F., Uchida, N., and Mori, K. (1999). Synchronized oscillarory discharge of mitral/tufted cells with different molecular receptive ranges in the rabbit olfactory bulb. J. Neurophysiol. *82*, 1786–1792.

Kellett, K.A., and Hooper, N.M. (2009). Prion protein and Alzheimer disease. Prion 3, 190–194.

Keshet, G.I., Bar-Peled, O., Yaffe, D., Nudel, U., and Gabizon, R. (2000). The cellular prion protein colocalizes with the dystroglycan complex in the brain. J. Neurochem. 75, 1889–1897.

Kim, C.K., Sakudo, A., Taniuchi, Y., Kang, C.B., Lee, D.C., Saeki, K., Matsumoto, Y., Sakaguchi, S., Itohara, S., and Onodera, T. (2007). Abnormal olfactory function caused by ectopic expression of Doppel in the olfactory bulb of prion protein-deficient mice. Intl. J. Mol. Med. *20*, 169–176.

Kramer, M.L., Kratzin, H.D., Schmidt, S., Bomer, A., Windl, O., Liemann, S., Hornemann, S., and Kretzschmar, H. (2001). Prion protein binds copper within the physiological concentration range. J. Biol. Chem. *276*, 16711–16719.

Kretzschmar, H.A., Tings, T., Madlung, A., Giese, A., and Herms, J. (2000). Function of PrP(C) as a copper-binding protein at the synapse. Arch. Virol. *16(Suppl.)*, 239–249.

Kristiansen, M., Deriziotis, P., Dimcheff, D.E., Jackson, G.S., Ovaa, H., Naumann, H., Clarke, A.R., Van Leeuwen, F.W., Menendez-Benito, V., Dantuma, N.P., *et al.* (2007). Disease-associated prion oligomers inhibit the 26S proteasome. Mol. Cell *26*, 175–188.

Kubosaki, A., Nishimura-Nasu, Y., Nishimura, T., Yusa, S., Sakudo, A., Saeki, K., Matsumoto, Y., Itohara, S., and Onodera, T. (2003). Expression of normal cellular prion protein (PrPc) on T lymphocytes and effect of copper ion: analysis by wild-type and prion gene-deficient mice. Biochem. Biophys. Res. Commun. *307*, 810–813.

Kurschner, C., and Morgan, J.I. (1995). The cellular prion protein (PrP) selectively binds to Bcl-2 in the yeast two-hybrid system. Brain Res. Mol. Brain Res. *30*, 165–168.

Kuwahara, C., Takeuchi, A.M., Nishimura, T., Haraguchi, K., Kubosaki, A., Matsumoto, Y., Saeki, K., Matsumoto, Y., Yokoyama, T., Itohara, S., *et al.* (1999). Prions prevent neuronal cell-line death. Nature *400*, 225–226.

Lagier, S., Panzanelli, P., Russo, R.E., Nissant, A., Bathellier, B., Sassoe-Pongnetto, M., Fritschy, J.M., and Lledo, P.M. (2007). GABAergic inhibition at dendrodendritic synapse tunes gamma oscillations in the olfactory bulb. Proc. Natl. Acad. Sci. U.S.A. *104*, 7259–7264.

Lassle, M., Blatch, G.L., Kundra, V., Takatori, T., and Zetter, B.R. (1997). Stress-inducible, murine protein mSTI1. Characterization of binding domains for heat shock proteins and *in vitro* phosphorylation by different kinase. J. Biol. Chem. *272*, 1876–1884.

Lauren, J., Gimble, D.A., Nygaard, H.B., Gilbert, J.W., and Strittmatter, S.M. (2009). Cellular prion protein mediates impairment of synaptic plasticity by amyloid-beta ologomers. Nature *457*, 1128–1132.

Le Pichon, C.E., Valley, M.T., Polymenidou, M., Chesler, A.T., Sagdullaev, B.T., Aguzzi, A., and Firestein, S. (2009). Olfactory behavior and physiology are disrupted in prion ptotein knockout mice. Nat. Neurosci. *12*, 60–69.

Li, A., Christensen, H.M., Stewart, L.R., Roth, K.A., Chiesa, R., and Harris, D.A. (2007). Neonatal lethalty in transgenic mice expressing prion protein with a deletion of residues 105–125. EMBO J. *26*, 548–558.

Linden, R., Martins, V.R., Prado, M.A., Cammarota, M., Izquierdo, I., and Brentani, R.R. (2008). Physiology of the prion protein. Physiol. Rev. *88*, 673–728.

Linden, R., Martins, V.R., and Prado, M.A. (2009). Prion protein. UCSD-Nature Molecules Pages, 29.

Liu, Y.H., Han, Y.L., Song, J., Wang, Y., Zhou, W., Zhang, B.Y., Tian, C., Li, C.P., Han, J., and Dong, X.P. (2010). Interaction between various 14–3–3beta segments and PrP *in vitro*. Zhonghua Shi Yan He Lin Chuang Bing Du Xue Za Zhi *24*, 165–167.

Lledo, P.M., and Lagier, S. (2006). Adjusting neurophysiological computation in the adult olfactory bulb. Semin. Cell Dev. Biol. *17*, 443–453.

Lloyd, S.E., Grizenkova, J., Pota, H., and Collinge, J. (2009). Shadoo (Sprn) and prion disease in incubation time in mice. Mamm. Genome *20*, 367–374.

Lopes, M.H., Hajj, G.N., Muras, A.G., Mancini, G.L., Castro, R.M., Ribeiro, K.C., Brentani, R.R., Linden, R., and Martins, V.R. (2005). Interaction of cellular prion and stress-inducible protein 1 promotes neuritogenesis and neuroprotection by distinct signaling pathways. J. Neurosci. *25*, 11330–11339.

Ma, J., Wollmann, R., and Lindquist, S. (2002). Neurotoxicity and neurodegeneration when PrP accumulates in the cytosol. Science *298*, 1781–1785.

McKinley, M.P., Taraboulos, A., Kenaga, L., Serban, D., Stieber, A., DeArmond, S.J., Prusiner, S.B., and Gonatas, N. (1991). Ultrastructural localization of kscrapie prion protein in cytoplasmic vesicles of infected cultured cells. Lab. Invest. *65*, 622–630.

McMahon, H.E., Mange, A., Nishida, N., Creminon, C., Casanova, D., and Lehmann, S. (2001). Cleavage of the amino terminus of the prion protein by reactive oxygen species. J. Biol. Chem. *276*, 2286–2291.

Mallucci, G.R., Ratté, S., Asante, E.A., Linehan, J., Gowland, I., Jefferys, J.G., and Collinge, J. (2002). Postnatal knock out of prion protein alters hippocampal CA1 properties, but does not result in neurodegeneration. EMBO J. *21*, 202–210.

Mallucci, G., Dickenson, A., Linehan, J., Klohn, P.C., Brander, S., and Collinge, J. (2003). Depleting neonatal PrP in prion infection prevents disease and reverse spongiosis. Science *302*, 871–874.

Mallucci, G.R., White, M.D., Farmer, M., Dickenson, A., Khatun, H., Powell, A.D., Brander, S., Jefferys, J.G., and Collinge, J. (2007). Targeting cellular prion protein reverses early congnitive deficits and neurophysiological dysfunction in prion-infected mice. Neuron *53*, 325–335.

Mange, A., Beranger, F., Peoc'h, K., Onodera, T., Frobert, Y., and Lehmann, S. (2004). Alpfa- and beta-cleavages of the amino-terminus of the cellular prion protein. Biol. Cell *96*, 125–132.

Manson, J., West, J.D., Thomson, V., McBride, P., Kaufman, M.H., and Hope, J. (1992). The prion protein gene: a role in mouse embryogenesis? Development *115*, 117–122.

Manson, J.C., Clarke, A.R., Hooper, M.L., Aitchinson, L., McConnell, I., and Hope, J. (1994). 129/Ola mice carrying a null mutation in PrP that abolishes mRNA productions are developmentally normal. Mol. Neurobiol. *8*, 121–127.

Martins, V.R., Graner, E., Garcia-Abreu, J., de Souza, S.J., Mercadante, A.F., Veiga, S.S., Zanata, S.M., Neto, V.M., and Brentani, R.R. (1997). Complementary hydropathy identifies a cellular prion protein receptor. Nat Med. *3*, 1376–1382.

Mattei, V., Garofalo, T., Misasi, R., Circella, A., Manganelli, V., Lucania, G., Pavan, A., and Sorice, M. (2004). Prion protein is a component of the multimolecular signaling complex involved in T cell activation. FEBS Lett. *560*, 14–18.

Martins, V.R., Beraldo, F.H., Hajj, G.N., Lopes, M.H., Lee, S.K., Prado, M.M., and Linden, R. (2010). Prion protein: orchestrating neurotrophic activities. Curr. Issues Mol. Biol, *12*, 63–86.

Meggio, F., Negro, A., Sarno, S., Ruzzene, M., Bertoli, A., Sorgato, M.C., and Pinna, L.A. (2000). Bovine prion protein as a modulator of protein kinase CK2. Biochem. J. *352*, 191–196.

Miele, G., Alejo Blanco, A.R., Baybutt, H., Horvat, S., Manson, J., and Clinton, M. (2003). Embryonic activation and developmental expression of the murine prion protein gene. Gene Expression *11*, 1–12.

Miesbauer, M., Bamme, T., Riemer, C., Oidtmann, B., Winklhofer, K.F., Baier, M., and Tatzelt, J. (2006). Prion protein-related proteins from zebrafish are complex glycosylated and contain a glycosylphpsphatidylinositol anchor. Biochem. Biophys. Res. Commun. *341*, 218–224.

Miura, T., Hori-i, A., Mototani, H., and Takeuchi, H. (1999). Raman spectroscopic study on the copper (II) binding mode of prion octapeptide and its pH dependence. Biochemistry *38*, 11560–11569.

Miyazawa, K., and Manualidis, L. (2010). Agent-specific Shadoo responses in transmissible encephalopathies. J. Neuroimmune Pharmacol. *5*, 155–163.

Moore, R.C., Lee, I.Y., Silverman, G.L., Harrison, P.M., Strome, R., Heinrich, C., Karunaratne, A., Pasternak, S.H., Chishti, M.A., Liang, Y., *et al.* (1999). Ataxia in prion protein (PrP)-deficient mice is associated with up-regulation of the novel PrP-like protein doppel. J. Mol. Biol. *292*, 797–817.

Moreno, J.A., Radford, H., Peretti, D., Steinert, J.R., Verity, N., Martin, M.G., Halliday, M., Morgan, J., Dinadale, D., Ortori, C.A., *et al.* (2012). Sustained translational repression by eIF2α-P mediates prion neurodegeneration. Nature *485*, 507–511.

Mouillet-Richard, S., Ermonval, M., Chebassier, C., Laplanche, J.L., Lehmann, S., Launay, J.M., and Kellermann, O. (2000). Signal transduction through prion protein. Science *289*, 1925–1928.

Mouillet-Richard, S., Pietri, M., Schneider, B., Vidal, C., Mutel, V., Launay, J.M., and Kellermann, O. (2005). Modulation of serotonergic receptor signaling and cross-talk by prion protein. J. Biol. Chem. *280*, 4592–4601.

Nazor, K.E., Seward, T., and Telling, G.C. (2007). Motor behavior and neuropathological deficits in mice deficient for normal prion protein expression. Biochem. Biophys. Acta *1772*, 645–653.

Nieznanski, K., Nieznanska, H., Skowronek, K.J., Osiecka, K.M., and Stepkowski, D. (2005). Direct interaction between prion protein and tubulin. Biochem. Biophys. Res. Commun. *334*, 403–411.

Nishida, N., Tremblay, P., Sugimoto, T., Shigematsu, K., Shirabe, S., Petromilli, C., Erpel, S.P., Nakaoke, R., Atarashi, R., Houtani, T., *et al.* (1999). A mouse prion protein transgene rescues mice deficient for the prion protein gene from Purkinje cell degeneration and demyelination. Lab. Invest. *79*, 689–697.

Nusser, Z., Kay, L.M., Laurent, G., Homanics, G.E., and Mody, I. (2001). Disruption of GABA(A) receptors on GABAergic interneurons leads to increased oscillatory power in the olfactory bulb network. J. Neurophysiol. *86*, 2823–2833.

Oesch, B., Teplow, D.B., Stahl, N., Serban, D., Hood, L.E., and Prusiner, S.B. (1990). Identification of cellular proteins binding to the scrapie prion protein. Biochemistry *29*, 5848–5855.

Pan, K.M., Baldwin, M., Nguyen, J., Gasset, M., Serban, A., Groth, D., Mehlhorn, I., Huang, Z., Fletterick, R.J., Cohen, F.E., *et al.* (1993). Conversion of alpfa-helix into beta-sheets features in the formation of the scrapie prion proteins. Proc. Natl. Acad. Sci. U.S.A. *90*, 10962–10966.

Park, J.S., Onodera, T., Nishimura, S., Thompson, R.F., and Itohara, S. (2006). Morecular evidence for two-stage learning and partial laterality in eyeblink conditioning of mice. Proc. Natl. Acad. Sci. U.S.A. *103*, 5549–5554.

Parkin, E.T., Watt, N.T., Hussain, I., Eckman, E.A., Eckman, C.B., Manson, J.C., Baybutt, H.N., Turner, A.J., and Hooper, N.M. (2007). Cellular prion protein regulates beta-secretase cleavage of the Alzheimer's amyloid precursor protein. Proc. Natl. Acad. Sci. U.S.A. *104*, 11062–11067.

Parkyn, C.J., Vermeulen, E.G., Mootoosamy, R.C., Sunyach, C., Lacobsen, C., Oxvig, C., Moestrup, S., Liu, Q., Bu, G., Jen, A., *et al.* (2008). LRP1 control biosynthetic and endocytic trafficking of neural prion protein. J. Cell Sci. *121*, 773–783.

Pattison, I.H., and Jebbett, J.N. (1973). Clinical and histological recovery from the scrapie-like spongiform encephalopathy produced in mice by feeding them with cuprizone. J. Pathol. *109*, 245–250.

Pauly, P.C., and Harris, D.A. (1998). Copper stimulates endocytosis of the prion protein. J. Biol. Chem. *273*, 33107–33110.

Peralta, O.A., Huckle, W.R., and Eyestona, W.H. (2011). Expression and knockdown of cellular prion protein (PrP^C) in differentiating mouse embryonic stem cells. Differentiation *81*, 68–77.

Premzl, M., Sangiorgio, L., Strumbo, B., Marshall Graves, J.A., Simonic, T., and Gready, J.E. (2003). Shadoo, a new protein highly conserved from fish to mammals and with similarity to prion protein. Gene *314*, 89–102.

Prusiner, S.B. (1997). Prion diseases and BSE crisis. Science *278*, 245–251.

Radvanivic, I., Braun, N., Giger, O.T., Mertz, K., Miele, G., Prinz, M., Navarro, B., and Aguzzi, A. (2005). Truncated prion protein and Doppel are myelinotoxic in the absence of oligodendrocytic PrPC. J. Neurosci. *25*, 4879–4888.

Rambold, A.S., Miesbauer, M., Rapaport, D., Bartke, T., Baier, M., Winklhofer, K.F., and Tatzelt, J. (2006). Association of Bcl-2 with misfolded prion protein is linked to the toxic potential of cytosolic PrP. Mol. Biol. Cell *17*, 3356–3368.

Rambold, A.S., Muller, V., Ron, U., Ben-Tal, N., Winklhofer, K.F., and Tatzelt, J. (2008). Stress-protective signaling of prion protein is corrupted by scrapie prions. EMBO J. *27*, 1974–1984.

Rieger, R., Edenhofer, F., Lasmezas, C.I., and Weiss, S. (1997). The human 37-kDa laminin receptor precursor interacts with the prion protein in eukaryotic cells. Nat. Med. *3*, 1383–1388.

Rossi, D., Cozzio, A., Flechsig, E., Klein, M.A., Rulicke, T., Aguzzi, A., and Weissmann, C. (2001). Onset of ataxia and Purkinje cell loss in PrP null mice inversely correlated with Dpl level in brain. EMBO J. *20*, 694–702.

Roucou, X., Giannopoulos, P.N., Zhang, Y., Jodoin, J., Goodyer, C.G., and LeBianc, A. (2005). Cellular prion protein inhibits proapoptotic Bax conformational change in human neurons and in breast carcinoma MCF-7 cells. Cell. Death Differ. *12*, 783–795.

Rutishauser, D., Mertz, K.D., Moos, R., Brunner, E., Rulicke, T., Calella, A.M., and Aguzzi, A. (2009). The comprehensive native interactome of a fully functional tagged prion protein. PLoS One *4*, e4446.

Sakaguchi, S., Katamine, S., Nishida, N., Moriuchi, R., Shigematsu, K., Sugimoto, T., Nakatani, A., Kataoka, Y., Houtani, T., Shirabe, S., *et al.* (1996). Loss of cerebellar Purkinje cells in aged mice homozygous for a disrupted PrP gene. Nature *380*, 528–531.

Sakthivelu, V., Seidel, R.P., Winklhofer, K.F., and Tatzelt, J. (2011). Conserved stress-protective activity between prion protein and Shadoo. J. Biol. Chem. *286*, 8901–8908.

Sakudo, A., Lee, D.C., Nishimura, T., Li, S., Tsuji, S., Nakamura, T., Matsumoto, Y., Saeki, K., Itohara, S., Ikuta, K., *et al.* (2005). Octapeptide repeat region and N-terminal half of hydrophobic region of prion protein (PrP) mediates PrP-dependent activation of superoxide dusmutase. Biochem. Biophys. Res. Commun. *326*, 600–606.

Sakudo, A., Onodera, T., Suganuma, Y., Kobayashi, T., Saeki, K., and Ikuta, K. (2006). Recent advances in clarifying prion protein functions using knockout mice and derived cell lines (Review). Mini Rev. Med. Chem. *6*, 589–601.

Santuccione, A., Sytnyk, V., Leshchyns'ka, I., and Schachner, M. (2005). Prion protein recruits its neuronal receptor NCAM to lipid rafts to activate p59fyn and to enhance neurite outgrowth. J. Cell Biol. *169*, 341–354.

Schmitt-Ulms, G., Legname, G., Baldwin, M.A., Ball, H.L., Bradon, N., Bosque, P.J., Crossin, K.L., Edelman, G.M., DeArmond, S.J., Cohen, F.E., *et al.* (2001). Binding of neural cell adhesion molecules (N-CAMs) to the cellular prion protein. J. Mol. Biol. *314*, 1209–1225.

Schneider, B., Mutel, V., Pietri, M., Ermonval, M., Mouillet-Richard, S., and Kellermann, O. (2003). NADPH oxidase are extracellular regulated kinase 1/2 are targets of prion protein signaling in neuronal and nonneuronal cells. Proc. Natl. Acad. Sci. U.S.A. *100*, 13326–13331.

Schoppa, N.E. (2006). Synchronization of olfactory bulb mitral cells by precisely timed inhibitory inputs. Neuron *49*, 271–283.

Shepherd, G.M. (2003). The Synaptic Organization of Brain (Oxford University Press, New York).

Shmerling, D., Hegyi, I., Fischer, M., Blattler, T., Brandner, S., Gotz, J., Bulicke, T., Flechsig, E., Cozzio, A., von Mering, C., *et al.* (1998). Expression of amino-terminally truncated PrP in mouse leading to ataxia and specific cerebellar lesions. Cell 93, 203–214.

Shyu, W.C., Chen, C.P., Saeki, K., Kubosaki, A., Matsumoto, Y., Onodera, T., Ding, D.C., Chiang, M.F., Lee, Y.J., Lin, S.Z., *et al.* (2005). Hypoglycemia enhances the expression of prion protein and heat-shock protein 70 in a mouse neuroblastoma cell line. J. Neurosci. Res. *80*, 887–894.

Solforosi, L., Criado, J.R., McGavern, D.B., Wirz, S., Sanchez-Alavez, M., Sugama, S., DeGiorgio, L.A., Volpe, B.T., Wiseman, E., Abalos, G., *et al.* (2004). Cross-linking cellular prion protein triggers neuronal apaotosis *in vivo*. Science *303*, 1514–1516.

Spielhaupter, C., and Schatzl, H.M. (2001). PrP^{C} directly interacts with protein involved in signaling pathways. J. Biol. Chem. *276*, 44604–44612.

Steele, A.D., Emsley, J.G., Ozdinler, P.H., Lidquist, S., and Macklis, J. (2006). Prion protein (PrPC) positively regulates neural precursor proliferation during developmental and adult mammalian neurogenesis. Proc. Natl. Acad. Sci. U.S.A. *103*, 3416–3421.

Steele, A.D., Lindquist, S., and Aguzzi, A. (2007). The prion protein knockout mouse: a phenotype under challenge. Prion *1*, 83–93.

Stopfer, M. (2007). Olfactory processing: massive convergence onto sparse codes. Curr. Biol. *17*, R363–364.

Strom, A., Wang, G.S., Picketts, D.J., Reimer, R., Stuke, A.W., and Scott, F.W. (2011). Cellular prion protein localizes to the nucleus of endocrine and neuronal cells and interacts with structural chromatin components. Eur. J. Cell Biol. *90*, 414–419.

Stys, P.K., You, H., and Zamponi, G.W. (2012). Copper-dependent regulation of NMDA receptors by cellular prion protein: implication for neurodegenerative disorders. J. Physiol. *590*, 1357–1368.

Sunyach, C., Cisse, M.A., da Costa, C.A., Vincent, B., and Checler, F. (2007). The C-terminal products of cellular prion protein processing, C1 and C2, exert distinct influence on p53-dependent staurosporine-induced caspase-3 activation. J. Biol. Chem. *282*, 1956–1963.

Taraboulos, A., Jendroska, K., Serban, D., Yang, S.L., DeArmond, S.J., and Prusiner, S.B. (1992). Regional mapping of prion proteins in brain. Proc. Natl. Acad. Sci. U.S.A. *15*, 7620–7624.

Taylor, D.R., and Hooper, N.M. (2007). The low-density lipoprotein receptor-related protein 1 (LRP1) mediates the endocytosis of the cellular prion protein. Biochem. J. *402*, 17–23.

Tompa, P., Fuxreiter, M., Oldfield, C.J., Simon, I., Dunker, A.K., and Uversky, V.N. (2009). Close encounters of the third kind: disordered domains and the interactions of proteins. Bioessays *31*, 328–335.

Tremblay, P., Bouzamondo-Bernstein, E., Heinrich, C., Prusiner, S.B., and DeArmond, S.J. (2007). Developmental expression of PrP in the post-implantation embryo. Brain Res. *30*, 60–67.

Tsayler, P., Harding, H.P., Ron, D., and Bertolotti, A. (2011). Selective inhibition of a regulatory subunit of protein phosphatase 1 restores proteostasis. Science 332, 91–94.

Urban, N.N. (2002). Lateral inhibition in the olfactory bulb and in olfaction. Physiol. Behavior 77, 607–612.

Vasallo, N., and Herms, J. (2003). Cellular prion protein function in copper homeostasis and redox signaling at the synapse. J. Neurochem. *86*, 538–544.

Vincent, B., Sunyach, C., Orzechowski, H.-D., St. George-Hyslop, P., and Checler, F. (2009). p53-Dependent transcriptional control of cellular prion by presenilins. J. Neurosci. *29*, 6752–6760.

Walmsley, A.R., Watt, N.T., Taylor, D.R., Perera, W.S., and Hooper, N.M. (2009). Alpha-cleavage of the prion protein occurs in a late compartment of the secretory pathway and is independent of lipid rafts. Mol. Cell. Neurosci. *40*, 242–248.

Wang, H.Y., Stucky, A., Liu, J., Shen, C., Trocme-Thibierge, C., and Morain, P. (2009). Dissociating beta-amyloid from alpfa 7 nicotinic acetylcholine and NMDA receptor function in Alzheimer's disease brain. J. Neurosci. *29*, 10961–10973.

Wang, H., Wan, J., Wang, W., Wang, D., Li, S., Liao, P., Hao, Z., Wu, S., Xu, J., Li, N., Ouyang, H., and Gao, H. (2011). Overexpression of Shadoo protein in transgenic mice does not impact the pathogenesis of scrapie. Neurosci. Lett. *496*, 1–4.

Wang, X., Wang, F., Arterburn, L., Wollmann, R., and Ma, J. (2006). The interaction between cytoplasmic prion protein and the hydrophobic lipid core of membrane correlates with neurotoxicity. J. Biol. Chem. *281*, 13559–13565.

Watarai, M., Kim, S., Erdenebaatar, J., Makino, S., Horiuchi, M., Shirahata, T., Sakaguchi, S., and Katamine, S. (2003). Cellular prion protein promotes Brucella infection into macrophages. J. Exp. Med. *198*, 5–17.

Watt, N.T., and Hooper, N.M. (2005). Reactive oxygen species (ROS)-mediated beta-cleavage of the prion protein in the mechanism of the cellular response to oxidative stress. Biochem. Soc. Trans. *33*, 1123–1125.

Watts, J.C., Drisaldi, B., Ng, V., Yang, J., Strome, B., Horne, P., Sy, M.S., Yoong, L., Young, R., Mastrangelo, P., *et al.* (2007). The CNS glycoprotein Shadoo has PrP(C)-like protective properties and displays reduced levels in prion infections. EMBO J. *26*, 4038–4050.

White, M.D., Farmer, M., Mirabile, I., Brander, S., Collinge, J., and Mallucci, G.R. (2008). Single treatment with RNAi against prion prptein rescues early neuronal dysfunction and prolongs survival in mice with prion disease. Proc. Natl. Acad. Sci. U.S.A. *105*, 10238–10243.

Winklhofer, K.F., Tatzelt, J., and Haass, C. (2008). The two faces of protein misfolding: gain-and loss-of-function in neurodegenerative diseases. EMBO J. 23, 336–349.

Yehiely, F., Bamborough, P., Da Costa, M., Perry, B.J., Thinakaran, G., Cohen, F.E., Carlson, G.A., and Prusiner, S.B. (1997). Identification of candidate proteins binding to prion protein. Neurobiol Dis. 3, 339–355.

Yokoi, M., Mori, K., and Nakanishi, S. (1995). Refreshment of odor molecule tuning by dendrodendritic synaptic inhibition in the olfactory bulb. Proc. Natl. Acad. Sci. U.S.A. *92*, 3371–3375.

Zafar, S., von Ahsen, N., Oellerich, M., Zerr, I., Schulz-Schaeffer, W.J., Armstrong, V.W., and Asif, A.R. (2011). Proteomics approach to identify the interacting partners of cellular prion protein and characterization of Rab7a interaction in neuronal cells. J. Proteome Res. *10*, 3123–3135.

Zanata, S.M., Lopes, M.H., Mercadante, A.F., Hajj, G.N., Chiarini, L.B., Nomizo, R., Freitas, A.R., Cabral, A.L., Lee, K.S., Juliano, M.A., *et al.* (2002). Stress- inducible protein 1 is a cell surface ligand for cellular prion that triggers neuroprotection. EMBO J. *21*, 3307–3316.

Zeng, F., Watt, N.T., Walmsley, A.R., and Hooper, N.M. (2003). Tethering the N-terminus of the prion protein compromises the cellular response to oxidative stress. J. Neurochem. *84*, 480–490.

Zhang, C.C., Steele, A.D., Lindquist, S.L., and Lodish, H.F. (2006). Prion protein is expressed on long-term repopulating hematopoietic stem cells and is important for their self-renewal. Proc. Natl. Acad. Sci. U.S.A. *103*, 2184.

Effect of Microglial Inflammation in Prion Disease

4

Yasuhisa Ano, Akikazu Sakudo and Takashi Onodera

Abstract

Prion diseases are a group of transmissible fatal neurodegenerative disorders. Neuropathological features of prion diseases include neuronal vacuolation, neuronal loss, astrogliosis and accumulation of activated microglial cells in affected brain areas. Recent studies have indicated that microglia may play a role in the pathogenesis of prion diseases. Chemokine receptor (CXCR3) expressed in microglia and microglial inflammasome activated by prion infection is important in the aetiology of CNS pathologies. The functions of microglia in this disease are discussed in this review.

Background

Prion diseases are a group of transmissible fatal neurodegenerative disorders of bovine spongiform encephalopathy (BSE) in cattle, Creutzfeldt–Jakob disease (CJD) in humans, scrapie in sheep, and chronic wasting disease (CWD) in deer and elk. They are caused by the conversion of cellular prion protein (PrP^{C}) into the pathological isoform (PrP^{Sc}) through conformational changes (Prusiner, 1998; Wechselberger, 2002). PrP^{Sc} is protease-resistant, and has a higher proportion of β-sheet structure in place of the normal α-helix structure. The accumulation of abnormal forms of prion protein (PrP^{Sc}) is important for developing the disease. Prion disease is neuropathologically characterized by neuronal vacuolation, neuronal loss, astrogliosis, and accumulation of activated microglial cells in affected brain areas (Ironside, 1998). Microglia, the resident macrophages of the central nervous system parenchyma, are exquisitely sensitive to pathological tissue alterations, undergoing morphological and phenotypic changes to adopt a so-called activated state and perform immunological functions in response to pathophysiological brain insults (Ransohoff *et al.*, 2009; Perry *et al.*, 2010). Many studies have demonstrated that the microglia have very diverse effector functions, in line with macrophage populations in other organs (Graeber, 2010). Mounting evidence also indicates that microglial activation contributes to neuronal damage in several neurodegenerative diseases including Alzheimer's disease, prion diseases, Parkinson's disease, multiple sclerosis, and Huntington's disease (Gonzalez *et al.*, 1999; Perry *et al.*, 2009). In prion diseases and other neurodegenerative disorders, microglia can become overactivated and release reactive oxygen species (ROS), nitric oxide (NO·), and cytokines, which can cause vascular damage in addition to neurodegeneration (Aguzzi *et al.*, 2006; Garção *et al.*, 2006; Block *et al.*, 2007). In this review, the roles of microglia in neurodegenerative and inflammatory process mediated by prion infection are discussed.

Microglial inflammation in the central nervous system

Microglia are the resident macrophages of the central nervous systems (CNS), in which they are ubiquitously distributed, accounting for approximately 10% of the adult brain cell population and representing the initial and primary immune response (Fig. 4.1A) (Kreutzberg, 1996).

Activated microglia are able to confer full immune effecter function, allowing them to eradicate the source of brain insult and to restore tissue integrity. This includes neuroprotective functions such as phagocytosis and cytotoxic effects via the release of pro-inflammatory mediators for swift removal of harmful pathogens (Fig. 4.1B) (van Rossum *et al.*, 2004; Neumann *et al.*, 2006; Imai *et al.*, 2007). Microglia activation occurs following exposure to CNS pathogen and detection of a variety of stimuli such as lipopolysaccharide (LPS), interferon-gamma (IFN-γ), amyloid-beta (Aβ) and other pro-inflammatory cytokines during injury and disease (Dheen *et al.*, 2007). Activated microglia can be identified and distinguished from their resting phenotype based on a combination of morphological and immunophenotypic changes. This includes a change from their typical ramified morphology to a reactive phenotype characterized by hypotrophy of the cell body, shortened and extensively branched processes (an amoeboid morphology), an1d significant up-regulation of cytoplasmic and membrane molecules (Ransohoff *et al.*, 2009).

To initiate innate immune responses, microglia enhance the expressions of toll-like receptors (TLR) (Bsibsi *et al.*, 2002) and multiple pro-inflammatory mediators such as tumour necrosis factor-alpha (TNF-α) (Floden *et al.*, 2005), interleukin (IL) 1 (Hartlage-Rubsamen *et al.*, 1999), and IL-6 (Suzumura *et al.*, 1996). The releases of various chemokines including macrophage inflammatory proteins 1α (MIP-1α) and MIP-1β (Takami *et al.*, 1997), monocyte chemoattractant protein-1 (MCP-1) (Babcock *et al.*, 2003), and also those involved in lymphocyte recruitment suggest that microglial activation is a process that precedes peripheral immune cell recruitment and that it is the first line of innate immunity in the CNS. Meanwhile, microglial cytotoxic functions are increased due to cytokine stimulation by other immune cells with the release of NO· (Banati *et al.*, 1993), such as superoxide (Chan *et al.*, 2007).

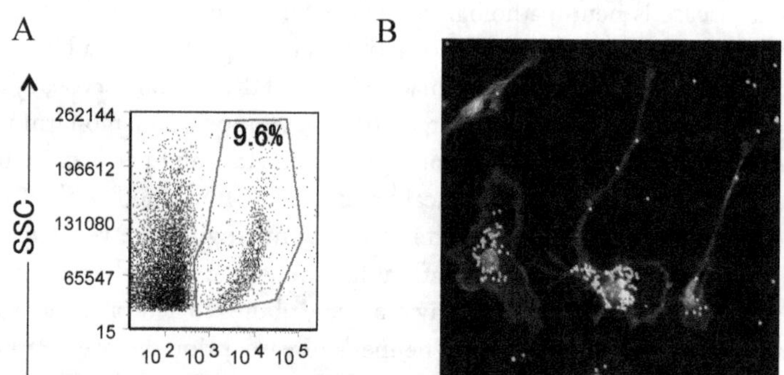

Figure 4.1 Population and phagocytosis of microglia. (A) The rate of microglia (CD11b-positive cells) in the brain cells is approximate 10%. (B) Microglia showed phagocytic activity and activated microglia released inflammatory cytokines. Green shows latex beads, red shows CD11b, and blue shows nuclei. Scale bar is 20µm. A colour version of this figure can be found in the plate section at the back of the book. SSC means side scatter.

Early post-mortem and histopathological investigations have reported the presence of large numbers of activated microglia in the CNS of patients with neurodegenerative disease including Alzheimer's disease, Parkinson's disease, multiple sclerosis, Huntington's disease, amyotrophic lateral sclerosis (ALS), and prion disease (Raine, 1994; Li *et al.*, 1996; Banati *et al.*, 1998; McGeer *et al.*, 1988a,b; Sitte *et al.*, 2001; Sapp *et al.*, 2001), although it remains inconclusive whether they play a role in pathogenesis or simply appear as a consequence of the disease process. The role of microglia as contributors to the progression of neurodegenerative disease was first proposed in Alzheimer's disease (Griffin *et al.*, 1989), where long-term intake of non-steroidal anti-inflammatory drugs (NSAIDs) was associated with a reduced risk of developing the disease (Etminan *et al.*, 2003; Vlad *et al.*, 2008). While a harmful role for microglia in neuroinflammation is a popular view, there is mounting evidence that points to the contrary, i.e. that microglia are in fact neuroprotective in these diseases (Hines *et al.*, 2009; Power *et al.*, 2009).

CXCR3 accelerates prion replication but prolongs survival times after prion infection

It has recently been reported that the chemokine receptor CXCR3 prolongs the survival periods in prion infection (Riemer *et al.*, 2008). In neurodegenerative disease, elevated chemokine expression levels have been observed in numerous pathologies of the brain. During prion infection in the CNS, the chemokines chemokine (C-C motif) ligand 2 (CCL2) (Felton *et al.*, 2005), CCL3 (Lu *et al.*, 2004; Riemer *et al.*, 2004), CCL5 (Marella *et al.*, 2004; Lee *et al.*, 2005), CCL6, CCL9, and CCL12 (Xiang *et al.*, 2004) have been found to be up-regulated. Moreover, induction of the chemokines CXCL9 (Schultz *et al.*, 2004), CXCL10, and CXCL13 (Riemer *et al.*, 2000) is seen at the early, asymptomatic stages of scrapie infection and is sustained at high levels throughout the disease process, possibly indicating an involvement in disease progression. In the periphery, these two groups of chemokines are potent chemoattractants for T cells (Klein *et al.*, 1997) and B cells (Lewicki *et al.*, 2003).

The chemokine receptor CXCR3 is widely expressed in brain tissue and has been found on astrocytes (Biber *et al.*, 2002; Flynn *et al.*, 2003), microglia (Priller *et al.*, 2006), neurons (Omari *et al.*, 2005), and oligodendrocytes (Horuk *et al.*, 2001). Established CXCR3 ligands are CXCL9, CXCL10, and CXCL11. Other chemokines, namely CCL21 and CXCL13, which are regular ligands of receptor CCR7 and CXCR5, respectively, are also thought to recognize CXCR3 (Jenh *et al.*, 2001; Rappert *et al.*, 2002). CXCR3 has been shown *in vitro* and *in vivo* to govern migration but not proliferation of microglia (Flynn *et al.*, 2003; Rappert *et al.*, 2004). In these studies, CXCR3-deficient mice infected with scrapie agents were characterized to determine the consequences of the impairment of microglial migration on disease development. CXCR3-deficient mice showed significantly prolonged survival time of up to 30 days on average. However, they displayed accelerated accumulation of misfolded proteinase K-resistant prion protein (PrP^{Sc}) and 20 times higher infectious prion titres than wild-type mice at the asymptomatic stage of the disease. In CXCR3-deficient animals, microglia activation was found to be reduced and quantitative analysis of gliosis-associated gene expression alterations demonstrated reduction in the number of proinflammatory factors (Riemer *et al.*, 2008). These findings suggested that inflammatory glial responses act in concern with PrP^{Sc} in disease development. CXCR3 is crucial for the recruitment of microglia to the inflammatory site of PrP^{Sc} deposition.

Neurotoxic prion peptide induces IL-1β production in microglia via inflammasome

The PrP fragment 106–126 (PrP106–126) is often used as PrP^{Sc} neurotoxic peptide in the research of prion disease, because PrP106–126 possesses similar physiochemical and pathological properties to PrP^{Sc}. PrP106–126 forms amyloid fibrils with a high β-sheet content, shows partial protease K resistance, and is neurotoxic *in vitro* (Forloni *et al.*, 1993; Selvaggini *et al.*, 1993; Henriques *et al.*, 2008).

Extensive research has indicated that the accumulation of aggregated PrP^{Sc} leads to activation of microglia, which in turn release superoxide, chemotactic factors, pro-inflammatory cytokines, and neurotoxic factors (Giese *et al.*, 1998; Marella *et al.*, 2004; Rock *et al.*, 2004). In addition, several studies have shown that multiple cytokines and chemokines such as IL-1β, TNF-α, and CCL3 were up-regulated in the brain from prion-infected mice (Tribouillard-Tanvier *et al.*, 2009). IL-1β plays a crucial role in the regulation of immune and inflammatory responses. It is produced as the inactive precursor pro-IL-1β in the cytosol, and a variety of stimuli lead to higher expression of pro-IL-1β (Dinarello, 2007). Pro-IL-1β is cleaved by caspase-1 into IL-1β, the active mature form. It has recently been reported that neurotoxic prions active mouse microglia and lead to IL-1β production (Peyrin *et al.*, 1999; Garcao *et al.*, 2006; Crozet *et al.*, 2008; Yang *et al.*, 2008). Shi *et al.* (2012) reported that PrP106–126 leads to the formation of NALP3 inflammasome in activated microglia. Inflammasome is a cytosolic protein complex that serves as a platform for activating the pro-inflammatory cytokines IL-1β and IL-18 via caspase-1 cleavage (Lamkanfi *et al.*, 2009). The inflammasome plays an important role in innate immunity and is involved in inflammatory disorders. NALP3, one of the most widely researched inflammasomes, consists of NACHT, LRR, and PYD domains-containing proteins, and is a well known as member NOD-like receptor family (Tschopp *et al.*, 2003; Mariathasan *et al.*, 2007). In this research, primary microglia cells from neonatal mice were primed with LPS and treated with PrP106–126. PrP106–126 activates caspase-1 and induces IL-1β release in the LPS-primed microglia. The expression of NALP3 and ASC was also up-regulated by PrP106–126 stimulation, indicating that neurotoxic prion peptide induces IL-1β via NALP3 inflammasome systems. It appears that inflammasome is a crucial mediator of severe inflammation and neuronal damage induced by microglia infected with prion.

It has recently been reported that CD36 plays an important role in microglial activation and IL-1β production triggered by PrP106–126 stimulation (Kouadir *et al.*, 2012). Blocking CD36 receptor reduces the microglial activation, i.e. the enhanced production of IL-1β, TNF-α and IL-6 associated with PrP106–126 treatment. The relationship of CD36 and inflammasome has not yet been elucidated; however, CD36 may be an important player in the inflammation caused by prion disease.

It has also been reported that NALP3 inflammasome is involved in the innate immune response to Aβ (Halle *et al.*, 2008). It is widely accepted that the extracellular accumulation of Aβ in senile plaques is a principal event in the pathogenesis of Alzheimer's disease (Weiner *et al.*, 2006; Meyer-Luehmann *et al.*, 2008). Microglia and invading bone marrow-derived mononuclear phagocytes are central to the initiation and progression of this disease. Microglia are activated by and recruited to senile plaques, whereupon they phagocytose Aβ and secrete cytokines after activation (Simard *et al.*, 2006). Moreover systemic inhibition of inflammation or immunization against Aβ decreases plaque burden and delays disease onset (Schenk *et al.*, 1999; Weggen *et al.*, 2001). One prominent cytokine consistently found in diseased tissues at early stages is IL-1β, which has been detected in microglia cells surrounding Aβ plaques in patients with Alzheimer's disease and in animal models of this disease. Similarly to PrP106–126, IL-1β is released from activated microglia in Alzheimer's disease via NALP3 inflammasome. Amyloid protein including PrP^{Sc} and Aβ may activate the inflammasome via a common mechanism.

Antioxidant cellular prion protein might contribute to control inflammasome in microglia

Recently, it is reported that ROS are essential secondary messengers which trigger NLRP3/NALP3 inflammasome activation (Dostert *et al.*, 2008; Zhou *et al.*, 2010). ROS production by H_2O_2 such as superoxide and hydroxyl radical activates the inflammasome, while knockdown of TRX, a cellular antioxidant protein, enhances IL-1β activation by silica, uric acid crystals, and asbestos (Dostert *et al.*, 2009). These findings suggest that oxidative stress could be sufficient to trigger NLRP3/NALP3 activation and the mechanism how NLRP3/NALP3 senses ROS. ATP-mediated ROS production has been shown to stimulate the PI3K pathway, and pharmacological inhibition of PI3K inhibits ATP-mediated caspase-1 activation, suggesting that PI3K may be involved in inflammasome activation downstream of ROS (Cruz *et al.*, 2007). It has also been reported that the non-steroidal anti-inflammatory drug aspirin inhibits the cytotoxicity of prion peptide PrP106–126 to neuronal cells associated with microglia activation *in vitro* (Yang *et al.*, 2008). ROS are generated mainly via the NADPH oxidase pathway and mitochondria. It has recently been reported that mitochondria control the activation of NLRP3 inflammasome and that the inflammasome activation is negatively regulated by autophagy and positively regulated by ROS. In that study, mitophagy/autophagy blockade led to accumulation of damaged, ROS-generating mitochondria, which in turn activated the NLRP3 inflammasome (Zhou *et al.*, 2010). The putative pathway involved in activation of inflammasome is depicted in Fig. 4.2.

Oxidation has been shown to be increased by prion infection. In brains from patients with Creutzfeldt–Jakob disease and from Syrian hamsters affected by scrapie, the amounts of glutamic and aminoadipic semialdehydes (products of metal-catalysed oxidation), malondialdehydelysine (a product of lipoxidation), N-epsilon-carboxyethyllysine (a product of glycoxidation), and N-epsilon-carboxymethyllysine (generated by lipoxidation and glycoxidation) were increased. The conversion of PrP^C into PrP^{Sc} was accompanied by alterations in fatty acid composition and increased phosphorylation of ERK(1/2) and p38, protein kinases known to respond to increased levels of ROS (Pamplona *et al.*, 2008).

In the CNS, PrP^C a cell-surface glycoprotein, is expressed mainly in neurons and also in glial cells, and it is an important factor in neurodegenerative prion diseases including bovine spongiform encephalopathy BSE, scrapie, and CJD. Kuwahara *et al.* (1999) reported that PrP^C-deficient neuronal cells die via apoptosis in serum-free medium, which indicates that PrP^C protects against oxidative stress under conditions of serum. Sakudo *et al.* (2005) revealed that PrP^C has superoxide dismutase activity and protects neurons from oxidative stress. Antioxidant PrP^C may contribute to suppress inflammasome activation. The relationship between PrP^C and inflammasome remains to be fully elucidated.

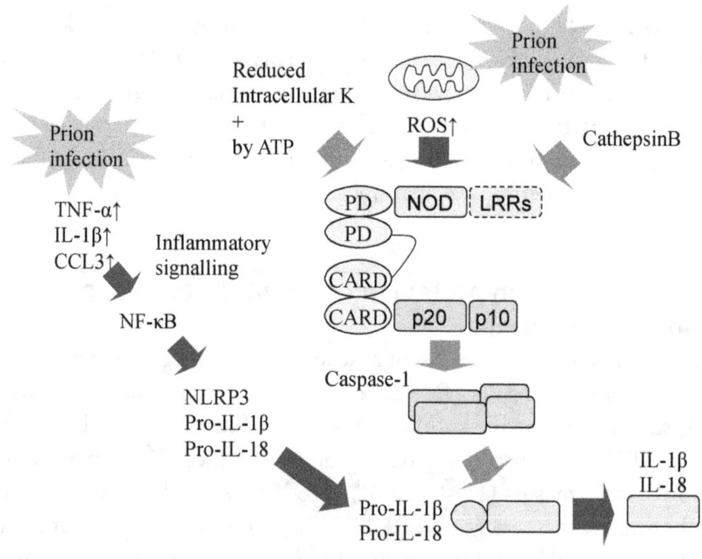

Figure 4.2 The schematic pathway of NLRP3 Inflammasome activation. IL-1β is produced via NLRP3 inflammasome activation. NLRP activation is associated with ROS production in mitochondria. A colour version of this figure can be found in the plate section at the back of the book.

Conclusion

In this review, the inflammatory reactions of microglia were examined in detail. Inflammatory responses have an important effect on progression of neurodegenerative disorders, but many aspects of the phenomena are not clear. The relationship between neurodegenerative disorders and inflammasome has been elucidated in the recent reports. The regulation of chronic inflammation in the brain is beneficial for to the treatment of neurodegenerative diseases including Alzheimer's disease, Parkinson's disease, Huntington's disease and prion disease.

References

Aguzzi, A., and Heikenwalder, M. (2006). Pathogenesis of prion diseases: current status and future outlook. Nat. Rev. Microbiol. *4*, 765–775.

Babcock, A.A., Kuziel, W.A., Rivest, S., and Owens, T. (2003). Chemokine expression by glial cells directs leukocytes to sites of axonal injury in the CNS. J. Neurosci. *23*, 7922–7930.

Banati, R.B., Gehrmann, J., Schubert, P., and Kreutzberg, G.W. (1993). Cytotoxicity of microglia. Glia 7, 111–118.

Banati, R.B., Daniel, S.E., and Blunt, S.B. (1998). Glial pathology but absence of apoptotic nigral neurons in long-standing Parkinson's disease. Mov. Disord. *13*, 221–227.

Biber, K., Dijkstra, I., Trebst, C., De Groot, C.J., Ransohoff, R.M., and Boddeke, H.W. (2002). Functional expression of CXCR3 in cultured mouse and human astrocytes and microglia. Neuroscience *112*, 487–497.

Block, M.L., Zecca, L., and Hong, J.S. (2007). Microglia-mediated neurotoxicity: uncovering the molecular mechanisms. Nat. Rev. Neurosci. *8*, 57–69.

Bsibsi, M., Ravid, R., Gveric, D., and van Noort, J.M. (2002). Broad expression of Toll-like receptors in the human central nervous systems. J. Neuropathol. Exp. Neurol. *61*, 1013–1021.

Chan, W.Y., Kohsaka, S., and Rezaie, P. (2007). The origin and cell linage of microglia: new concepts. Brain Res. Rev. *53*, 344–354.

Crozet, C., Beranger, F., and Lehmann, S. (2008). Cellular pathogenesis in prion diseases. Vet. Res. *39*, 44.

Cruz, C.M., Rinna, A., Forman, H.J., Ventura, A.L., Persechini, P.M., and Ojcius, D.M. (2007). ATP activates a reactive oxygen species-dependent oxidative stress response and secretion of proinflammatory cytokines in macrophages. J. Biol. Chem. *282*, 2871–2879.

Dheen, S.T., Kaur, C., and Ling, E.A. (2007). Microglial activation and its implications in the brain diseases. Curr. Med. Chem. *14*, 1189–1197.

Dinarello, C.A. (2007). A signal for the caspase-1 inflammasome free of TLR. Immunity *26*, 383–385.

Dostert, C., Pétrilli, V., Van Bruggen, R., Steele, C., Mossman, B.T., and Tschopp, J. (2008). Innate immune activation through Nalp3 inflammasome sensing of asbestos and silica. Science *320*, 674–677.

Dostert, C., Guarda, G., Romero, J.F., Menu, P., Gross, O., Tardivel, A., Suva, M.L., Stehle, J.C., Kopf, M., Stamenkovic, I., *et al.* (2009). Malarial hemozoin is a Nalp3 inflammasome activating danger signal. PLoS One *4*, e6510.

Etminan, M., Gill, S., and Samii, A. (2003). Effect of non-steroidal anti-inflammatory drugs on risk of Alzheimer's disease: systematic review and meta-analysis of observational studies. BMJ *327*, 128.

Felton, L.M., Cunningham, C., Rankine, E.L., Waters, S., Boche, D., and Perry, V.H. (2005). MCP-1 and murine prion disease: separation of early behavioural dysfunction from overt clinical disease. Neurobiol. Dis. *20*, 283–295.

Floden, A.M., Li, S., and Combs, C.K. (2005). Beta-amyloid-stimulated microglia induce neuron death via synerigistic stimulation of tumor necrosis factor alpha and NMDA receptors. J. Neurosci. *25*, 2566–2575.

Flynn, G., Maru, S., Loughlin, J., Romero, I.A., and Male, D. (2003). Regulation of chemokine receptor expression in human microglia and astrocytes. J. Neuroimmunol. *136*, 84–93.

Forloni, G., Angeretti, N., Chiesa, R., Monzani, E., Salmona, M., Bugiani, O., and Tagliavini, F. (1993). Neurotoxicity of a prion protein fragment. Nature *362*, 543–546.

Garcao, P., Oliveira, C.R., and Agostinho, P. (2006). Comparative study of microglia activation induced by amyloid-beta and prion peptides: role in neurodegeneration. J. Neurosci. Res. *84*, 182–193.

Giese, A., Brown, D.R., Groschup, M.H., Feldmann, C., Haist, I., and Kretzschmar, H.A. (1998). Role of microglia in neuronal cell death in prion disease. Brain Pathol. *8*, 449–457.

González-Scarano, F., and Baltuch, G. (1999). Microglia as mediators of inflammatory and degenerative diseases. Ann. Rev. Neurosci. *22*, 219–240.

Graeber, M.B. (2010). Changing face of microglia. Science *330*, 783–788.

Griffin, W.S., Stanley, L.C., Ling, C., White, L., MacLeod, V., Perrot, L.J., White, C.L. 3rd, and Araoz, C. (1989). Brain interleukin 1 and S-100 immunoreactivity are evaluated in Down syndrome and Alzheimer's disease. Proc. Natl. Acad. Sci. U.S.A. *86*, 7611–7615.

Halle, A., Hornung, V., Petzold, G.C., Stewart, C.R., Monks, B.G., Reinheckel, T., Fitzgerald, K.A., Latz, E., Moore, K.J., and Golenbock, D.T. (2008). The NALP3 inflammasome is involved in the innate immune response to amyloid-beta. Nat. Immunol. *9*, 857–865.

Hartlage-Rubsamen, M., Lemke, R., and Schiliebs, R. (1999). Interleukin-1 beta, inducible nitric oxide synthase, and nuclear factor-kappaB are induced in morphologically distinct microglia after rat hippocampal lipopolysaccharide/interferon-gamma injection. J. Neurosci. Res. *57*, 388–398.

Henriques, S.T., Pattenden, L.K., Aguilar, M.I., and Castanho, M.A.R.B. (2008). PrP (106–126) does not interact with membranes under physiological conditions. Biophys. J. *95*, 1877–1889.

Hines, D.J., Hines, R.M., Mulligan, S.J., and Macvicar, B.A. (2009). Microglia processes block the spread of damage in the brain and require functional chloride channels. Glia *57*, 1610–1618.

Horuk, R. (2001). Chemokine receptors. Cytokine Growth Factor Rev. *12*, 313–335.

Imai, F., Suzuki, H., Oda, J., Ninomiya, T., Ono, K., Sano, H., and Sawada, M. (2007). Neuroprotective effect of exogenous microglia in global brain ischemia. J. Cereb. Blood Flow *27*, 488–500.

Ironside, J.W. (1998). Prion diseases in man. J. Pathol. *186*, 227–234.

Jenh, C.H., Cox, M.A., Hipkin, W., Lu, T., Pugliese-Sivo, C., Gonsiorek, W., Chou, C.C., Narula, S.K., and Zavodny, P.J. (2001). Human B cell-attracting chemokine 1 (BCA-1; CXCL13) is an agonist for the human CXCR3 receptor. Cytokine *15*, 113–121.

Klein, M.A., Frigg, R., Flechsig, E., Raeber, A.J., Kalinke, U., Bluethmann, H., Bootz, F., Suter, M., Zinkernagel, R.M., and Aguzzi, A. (1997). A crucial role for B cells in neuroinvasive scrapie. Nature *390*, 687–690.

Kouadir, M., Yang, L., Tan, R., Shi, F., Lu, Y., Zhang, S., Yin, X., Zhou, X., and Zhao, D. (2012). CD36 participates in PrP(106–126)-induced activation of microglia. PLos One *7*, e30756.

Kreutzberg, G.W. (1996). Maicroglia: a sensor for pathological events in the CNS. Trends Neurosci. *19*, 312–318.

Kuwahara, C., Takeuchi, A.M., Nishimura, T., Haraguchi, K., Kubosaki, A., Matsumoto, T., Saeki, K., Matsumoto, T., Yokoyama, T., Itohara, S., *et al.* (1999). Prions prevent neuronal cell-line death. Nature *400*, 225–226.

Lamkanfi, M., and Dixit, V.M. (2009). The inflammasomes. PLoS Pathog. *5*, e1000510.

Lee, H.P., Jun, Y.C., Choi, J.K., Kim, J.I., Carp, R.I., and Kim. Y.S. (2005). The expression of RANTES and chemokine receptors in the brains of scrapie-infected mice. J. Neuroimmunol. *158*, 26–33.

Lewicki, H., Tishon, A., Homann, D., Mazarguil, H., Laval, F., Asensio, V.C., Campbell, I.L., DeArmond, S., Coon, B., Teng, C., *et al.* (2003). T cells infiltrate the brain in murine and human transmissible spongiform encephalopathies. J. Virol. *77*, 3799–3808.

Li, H., and Cuzner, M.L., and Newcombe, J. (1996). Microglial-derived macrophages in early multiple sclerosis plaques. Neuropathol. Appl. Neurobiol. *22*, 207–215.

Lu, Z.Y., Baker, C.A., and Manuelidis, L. (2004). New molecular markers of early and progressive CJD brain infection. J. Cell. Biochem. *93*, 644–652.

McGeer, P.L., Itagaki, S., Boyes, B.E., and McGeer, E.G. (1988a). Reactive microglia are positive for HLA-DR in the substantia nigra of Parkinson's and Alzheimer's disease brains. Neurology *38*, 1285–1291.

McGeer, P.L., Itagaki, S., and McGeer, E.G. (1988b). Expression of the histocompatibility glycoprotein HLA-DR in neurological disease. Acta. Neuropathol. *76*, 550–557.

Marella, M., and Chabry, J. (2004). Neurons and astrocytes respond to prion infection by inducing microglia recruitment. J. Neurosci. *24*, 620–627.

Mariathasan, S., and Monack, D.M. (2007). Inflammasome adaptors and sensors: intracellular regulators of infection and inflammation. Nat. Rev. Immunol. *7*, 31–40.

Meyer-Luehmann, M. (2008). Rapid appearance and local toxicity of amyloid-β plaques in a mouse model of Alzheimer's disease. Nature *451*, 720–724.

Neumann, J., Gunzer. M., Gutzeit, H.O., Ullrich, O., Reymann, K.G., and Dingkel, K. (2006). Microglia cells provide neuroprotection after ischemia. FASEB J. *20*, 714–716.

Omari, K.M., John, G.R., Sealfon, S.C., and Raine, C.S. (2005). CXC chemokine receptors on human oligodendrocytes: implications for multiple sclerosis. Brain *128*, 1003–1015.

Pamplona, R., Naudi, A., Gavin, R., Pastrana, M.A., Sajnani, G., Ilieva, E.V., Del Rio, J.A., Portero-Otin, M., Ferrer, I., and Requena, J.R. (2008). Increased oxidation, glycoxidation, and lipoxidation of brain proteins in prion disease. Free Radic. Biol. Med. *45*, 1159–1166.

Perry, V.H., Nicoll, J.A., and Holmes, C. (2010). Microglia in neurodegenerative disease. Nat. Rev. Neurol. *6*, 193–201.

Peyrin, J.M., Lasmézas, C.I., Haïk, S., Tagliavini, F., Salmona, M., Williams, A., Richie, D., Deslys, J.P., and Dormont, D. (1999). Microglial cells respond to amyloidogenic PrP peptide by the production of inflammatory cytokines. Neuroreport *10*, 723.

Power, J.H., and Blumbergs, P.C. (2009). Cellular glutathione peroxidase in human brain: cellular distribution, and its potential role in the degradation of Lewy bodies in Parkinson's disease and dementia with Lewy bodies. Acta. Neuropathol. *117*, 63–73.

Priller, J., Prinz, M., Heikenwalder, M., Zeller, N., Schwarz, P., Heppner, F.L., and Aguzzi, A. (2006). Early and rapid engraftment of bone marrow-derived microglia in scrapie. J. Neurosci. *26*, 11753–11762.

Prusiner, S.B. (1998). Prions. Proc. Natl. Acad. Sci. USA *95*, 13363–13383.

Raine C.S. (1994). Multiple sclerosis: immune system molecules expression in the central nervous system. J. Neuropathol. Exp. Neurol. *53*, 328–337.

Ransohoff, R.M., and Perry, V.H. (2009). Microglial physiology: unique stimuli, Specialized Responses. Annu. Rev. Immunol. *27*, 119–145.

Rappert, A., Bechmann, I., Pivneva, T., Mahlo, J., Biber, K., Nolte, C., Kovac, A.D., Gerard, C., Boddeke, H.W., Nitsch, R., *et al.* (2004). CXCR3-dependent microglial recruitment is essential for dendrite loss after brain lesion. J. Neurosci. *24*, 8500–8509.

Rappert, A., Biber, K., Nolte, C., Lipp, M., Schubel, A., Lu, B., Gerard, N.P., Gerard, C., Boddeke, H.W., and Kettenmann, H. (2002). Secondary lymphoid tissue chemokine (CCL21) activates CXCR3 to trigger a Cl current and chemotaxis in murine microglia. J. Immunol. *168*, 3221–3226.

Riemer, C., Queck, I., Simon, D., Kurth, R., and Baier. M. (2000). Identification of up-regulated genes in scrapie-infected brain tissue. J. Virol. *74*, 10245–10248.

Riemer, C., Neidhold, S., Burwinkel, M., Schwarz, A., Schultz, J., Kratzschmar, J., Monning, U., and Baier. M. (2004). Gene expression profiling of scrapie-infected brain tissue. Biochem. Biophys. Res. Commun. *323*, 556–564.

Riemer, C., Schultz, J., Burwinkel, M., Schwarz, A., Mok, S.W., Gültner, S., Bamme, T., Norley, S., van Landeghem, F., Lu, B., *et al.* (2008). Accelerated prion replication in, but prolonged survival times of, prion-infected CXCR3-/- mice. J. Virol. *82*, 12464–12471.

Rock, R.B., Gekker, G., Hu, S., Sheng, W.S., Cheeran, M., Lokensgard, J.R., and Peterson, P.K. (2004). Role of microglia in central nervous system infections. Clin. Microbiol. Rev. *17*, 942–964.

van Rossum, D., and Hanisch, U.K. (2004). Microglia. Metab. Brain Dis. *19*, 393–411.

Sakudo, A., Lee, D.C., Nishimura, T., Li, S., Tsuji, S., Nakamura, T., Matsumoto, Y., Saeki, K., Itohara, S., Ikuta, K., *et al.* (2005). Octapeptide repeat region and N-terminal half of hydrophobic region of prion protein (PrP) mediate PrP-dependent activation of superoxide dismutase. Biochem. Biophys. Res. Commun. *326*, 600–606.

Sapp, E., Kegel, K.B., Aronin, N., Hashikawa, T., Uchiyama, Y., Tohyama, K., Bhide, P.G., Vonsattel, J.P., and DiFiglia, M. (2001). Early and progressive accumulation of reactive microglia in the Huntington disease brain. J. Neuropathol. Exp. Neurol. *60*, 161–172.

Schenk, D., Barbour, R., Dunn, W., Gordon, G., Grajeda, H., Guido, T., Hu, K., Huang, J., Johnson-Wood, K., Khan, K., *et al.* (1999). Immunization with amyloid-beta attenuates Alzheimer-disease-like pathology in the PDAPP mouse. Nature *400*, 173–177.

Schultz, J., Schwarz, A., Neidhold, S., Burwinkel, M., Riemer, C., Simon, D., Kopf, M., Otto, M., and Baier, M. (2004). Role of interleukin-1 in prion disease associated astrocyte activation. Am. J. Pathol. *165*, 671–678.

Selvaggini, C., Degioia, L., Cantu, L., Ghibaudi, E., Diomede, L., Passerini, F., Forloni, G., Bugiani, O., Tagliavini, F., and Salmona, M. (1993). Molecular characteristics of a protease-resistant, amyloidogenic and neurotoxic peptide homologous to residues 106–126 of the prion protein. Biochem. Biophys. Res. Commun. *194*, 1380–1386.

Shi, F., Yang, L., Kouadir, M., Yang, Y., Wang, J., Zhou, X., Yin, X., and Zhao, D. (2012). The NALP3 inflammasome is involved in neurotoxic prion peptide-induced microglial activation. J. Neuroinflammation 9, 73.

Simard, A.R., Soulet, D., Gowing, G., Julien, J.P., and Rivest, S. (2006). Bone marrow-derived microglia play a critical role in restricting senile plaque formation in Alzheimer's disease. Neuron *49*, 489–502.

Sitte, H.H., Wanshitz, J., Budka, H., and Burger, M.L. (2001). Autoradiography with [3H]PK11195 of spinal tract degeneration in amyotrophic lateral sclerosis. Acta. Neuropathol. *101*, 75–78.

Suzumura, A., Sawada, M., and Marunouchi, T. (1996). Selective induction of interleukin-6 in mouse microglia by granulocyte–macrophage colony-stimulating factor. Brain Res. *713*, 192–198.

Takami, S., Nishikawa, H., Minami, M., Nishiyori, A., Sato, M., Akaike, A., and Satoh, M. (1997). Induction of macrophage inflammatory protein MIP-1 alpha mRNA on glial cells after focal cerebral ischemia in the rat. Neurosci. Lett. *227*, 173–176.

Tribouillard-Tanvier, D., Striebel, J.F., Peterson, K.E., and Chesebro, B. (2009). Analysis of protein levels of 24 cytokines in scrapie agent-infected brain and glial cell cultures from mice differing in prion protein expression levels. J. Virol. *83*, 11244.

Tschopp, J., Martinon, F., and Burns, K. (2003). NALPs: a novel protein family involved in inflammation. Nat. Rev. Mol. Cell. Biol. *4*, 95–104.

Vlad, S.C., Miller, D.R., Kowall, N.W., and Felson, D.T. (2008). Protective effects of NSAIDs on the development of Alzheimer's disease. Neurology *70*, 1672–1677.

Wechselberger, C., Wurm, S., Pfarr, W., and Höglinger, O. (2002). The physiological functions of prion protein. Exp. Cell Res. *281*, 1–8.

Weggen, S., Eriksen, J.L., Das, P., Sagi, S.A., Wang, R., Pietrzik, C.U., Findlay, K.A., Smith, T.E., Murphy, M.P., Bulter, T., *et al.* (2001). A subset of NSAIDs lower amyloidogenic Abeta42 independently of cyclooxygenase activity. Nature *414*, 212–216.

Weiner, H.L., and Frenkel, D. (2006). Immunology and immunotherapy of Alzheimer's disease. Nat. Rev. Immunol. *6*, 404–416.

Xiang, W., Windl, O., Wunsch, G., Dugas, M., Kohlmann, A., Dierkes, N., Westner, I.M., and Kretzschmar, H.A. (2004). Identification of differentially expressed genes in scrapie-infected mouse brains by using global gene expression technology. J. Virol. *78*, 11051–11060.

Yang, L., Zhou, X., Yang, J., Yin, X., Han, L., and Zhao, D. (2008). Aspirin inhibits cytotoxicity of prion peptide PrP106–126 to neuronal cells associated with microglia activation *in vitro*. J. Neuroimmunol. *199*, 10–17.

Zhou, R., Tardivel, A., Thorens, B., Choi, I., and Tschopp, J. (2010). Thioredoxin-interacting protein links oxidative stress to inflammasome activation. Nat. Immunol. *11*, 136–140.

Zhou, R., Yazdi, A.S., Menu, P., and Tschopp, J. (2011). A role for mitochondria in NLRP3 inflammasome activation. Nature *469*, 221–225.

Molecular Mechanisms of Prion Diseases

5

Hermann C. Altmeppen, Berta Puig, Susanne Krasemann, Clemens Falker, Frank Dohler and Markus Glatzel

Abstract

Prion diseases or transmissible spongiform encephalopathies are fatal neurodegenerative conditions occurring in humans and animals. Experimental data and neuropathological examinations show that prions (here defined as the agent responsible for transmissible spongiform encephalopathies) consist of a self-propagating isoform of the cellular prion protein. Nucleic acids are not required for propagation of prions. In the last years a number of questions regarding the mechanism of prion propagation and neurotoxicity as well as the spread of prions to and within the brain have been answered, yet essential pieces of information regarding the execution of cell death and cell-to-cell spread of prions remain to be elucidated.

Introduction

Transmissible spongiform encephalopathies (TSEs) or prion diseases are neurodegenerative disorders with a fatal disease course occurring both in humans and animals (Geissen *et al.*, 2007). Although neuropathological hallmarks such as spongiosis and deposition of abnormally folded prion protein are common in all prion diseases of humans and animals, their origin is diverse. In fact for humans, prion disease may arise as a sporadic disease or may also be of genetic or acquired nature, whereas in animals these diseases are either of sporadic or acquired origin (Glatzel and Aguzzi, 2001). Human prion diseases include sporadic, genetic or acquired forms of Creutzfeldt–Jakob disease (CJD) among others and the most prominent prion diseases of animals are bovine spongiform encephalopathy (BSE), scrapie of sheep (note the abbreviation 'Sc' for scrapie in the pathogenic protein isoform PrP^{Sc}), and chronic wasting disease of deer and elk (CWD). In acquired forms of the disease, transmission has occurred by exposure to infectious prions (Will, 2003). Interspecies transmission of prions is rare and impeded by a considerable species barrier; however a variant form of CJD in humans is almost certainly caused by exposure of humans towards BSE-prions.

Recent insights into the physiological role of the prion protein have helped to understand how misfolding and subsequent deposition of this protein associate with development of prion disease and it has become obvious that propagation of prions by templated misfolding lies at the basis of the pathophysiology of prion diseases (Colby and Prusiner, 2011). Better understanding of the cell biology of the prion protein has provided important clues on how aberrant signalling contributes to synaptic degeneration and neuronal death, the underlying cause of progressive dementia observed in prion disease. In this chapter we describe the molecular mechanisms underlying the development of prion diseases focussing on the role of the prion protein including its post-translational modifications in this process.

Prion diseases in humans and animals

The existence of a group of diseases with peculiar properties, now referred to as prion diseases, occurring both in humans and in animals, was realized centuries ago (one century for human prion diseases, two centuries for animal prion diseases).

For human prion diseases, the German neurologists Alfons Maria Jakob and Hans-Gerhard Creutzfeldt described a group of patients with rapidly progressing dementia and neurodegeneration in 1921 (Jakob, 1921). It is now obvious that human prion diseases present with a wide phenotypical spectrum. Besides the common forms of CJD, such as sporadic CJD, there are genetic variants, where the disease co-segregates with mutations in the gene encoding the prion protein (genetic CJD), and acquired forms (variant and iatrogenic CJD) (Ironside, 1998; Glatzel *et al.*, 2002; Geissen *et al.*, 2007). Acquired forms of CJD have occurred by (i) ritualistic cannibalism and consumption of dead relatives in the case of Kuru (Gajdusek and Reid, 1961), (ii) transmission of BSE-prions to humans in the case of variant CJD (Will *et al.*, 1996; Hill *et al.*, 1997), or (iii) by exposure to infectious prions during medical procedures in the case of iatrogenic CJD (Brown *et al.*, 2000). In acquired forms of CJD, the penetrance is incomplete (resulting in the fact that only a fraction of persons potentially exposed to infectious prions develop a prion disease) and incubation times from putative exposure towards onset of prion disease are exceedingly long, possibly surpassing decades (Collinge *et al.*, 2006). For vCJD, iatrogenic transmission of the disease by blood transfusion from subclinically infected vCJD carriers has occurred (Llewelyn *et al.*, 2004; Wroe *et al.*, 2006). The overall incidence of all forms of human prion disease is low, with an incidence of approximately two cases/million/year although higher incidence rates have been reported (Glatzel *et al.*, 2002; Stoeck *et al.*, 2008).

The first prion disease described in animals in the early eighteenth century was scrapie, a prion disease affecting sheep (Parry, 1960). The most prominent prion disease of animals is bovine spongiform encephalopathy (BSE) or 'Mad Cow disease' which became endemic in the UK during the late 1980s and early 1990s (Wells *et al.*, 1987). Drastic measures limiting dietary spread of the disease within cattle populations resulted in a progressive decline in the number of diagnosed cases of BSE (Wells *et al.*, 1987). Recently, atypical BSE types such as bovine amyloidotic spongiform encephalopathy (BASE), which may well represent sporadic form of BSE, have been described (Capobianco *et al.*, 2007).

Another relevant form of prion disease of animals is chronic wasting disease of deer and elk (CWD). This is a prion disease of unknown origin affecting captive and free-ranging cervids and occurs mainly in the United States (Sigurdson *et al.*, 2008).

All TSEs have a common pattern of pathological alterations found in the brain. These include spongiform vacuolation (spongiosis), neuronal loss, astrogliosis, microgliosis, and the deposition of abnormally folded prion protein. Deposition of abnormally folded prion protein occurs in a wide range of pattern including minute deposits unevenly distributed in cortical areas (synaptic pattern), as patchy deposits, intraneuronally or in the form of amyloid plaques and become obvious only upon immunolabelling with antibodies directed against the prion protein (Fig. 5.1). Although deposition of abnormally folded prion protein may also be found in organ systems distinct from the central nervous system (i.e. the

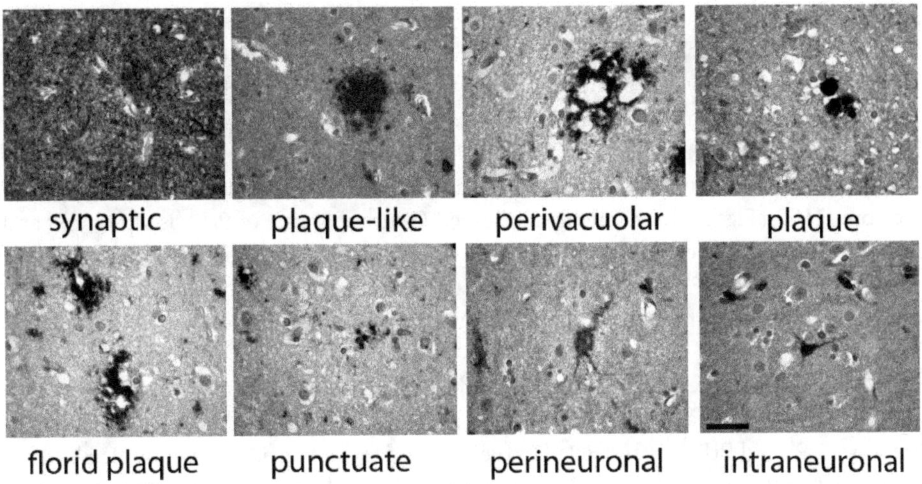

Figure 5.1 Deposition patterns of abnormally folded prion protein in the brain. Shown are representative examples of deposition patterns described for the deposition of abnormally folded prion protein. These only become visible upon immunolabelling with antibodies directed against the prion protein. Nomenclature of deposition patterns is indicated below histology. Scale bar represents 50μm. A colour version of this figure can be found in the plate section at the back of the book.

lymphoreticular system or the muscular system), extraneural tissues do not display obvious signs of histopathological damage (Haley *et al.*, 2009; Balkema-Buschmann *et al.*, 2011). In some naturally occurring prion diseases such as CWD, and in experimentally induced murine prion disease, prions can be found in bodily fluids or excretions such as saliva, faeces, urine, and blood (Seeger *et al.*, 2005; Haley *et al.*, 2009).

The prion protein

As discussed previously (Altmeppen *et al.*, 2012), the mature form of the cellular prion protein (PrP^C) is a membrane-anchored glycoprotein. PrP^C is discussed to fulfil several physiological functions (Aguzzi *et al.*, 2008; Linden *et al.*, 2008). These range from involvement in neuro-, synapto-, and neuritogenesis, (Graner *et al.*, 2000; Steele *et al.*, 2006) cell differentiation (Hajj *et al.*, 2007), cell adhesion (Mange *et al.*, 2002; Malaga-Trillo *et al.*, 2009), neuroprotection (Chiarini *et al.*, 2002; Rambold *et al.*, 2008), and copper-homeostasis (Brown, 2003), to receptor properties and participation in cellular signalling pathways. In signalling, PrP^C can itself be involved (Solforosi *et al.*, 2004; Mouillet-Richard *et al.*, 2007; Resenberger *et al.*, 2011a) or act as a regulatory cofactor (You *et al.*, 2012). In both situations, accessory molecules are required since PrP^C does not span the plasma membrane and is thus unable to transduce signals into the cytosol.

The cellular prion protein comprises 253 amino acids. The mature form of PrP^C contains 209 amino acids (murine PrP^C) with a molecular weight of 21 to 35 kDa depending on its glycosylation status (see below). In protein synthesis, during translocation of PrP^C into the endoplasmic reticulum, the secretory signal peptide of 22 amino acids is cleaved from the N-terminus of PrP^C, and the C-terminal signal sequence for addition of the glycosylphosphatidylinositol (GPI)-anchor to attach PrP^C to the outer leaflet of membranes is processed (Riek *et al.*, 1997; Biasini *et al.*, 2012). PrP^C contains two variably occupied N-glycosylation sites where complex-type oligosaccharides are added at positions 181 and 197 (murine PrP^C). Thus, in western blot analysis, three major bands are visible reflecting PrP^C with two, one, or no glyco-residues attached. The flexible N-terminal part contains the octameric repeat region (OR) and a neurotoxic domain (ND) in addition to the hydrophobic core (HC) (Fig. 5.2). The highly conserved C-terminal part of PrP^C is structured and contains three alpha-helices, two beta-sheets, loop domains, and a disulfide bond. Because of its GPI anchor, PrP^C is mainly located within cholesterol and sphingolipid-rich membrane microdomains, termed lipid rafts (Brugger *et al.*, 2004; Taylor and Hooper, 2006; Puig *et al.*, 2011).

PrP^C and PrP^{Sc} (see below under 'Misfolding of the prion protein') share the identical amino acid sequence and harbour identical post-translational modifications. Differences between PrP^C and PrP^{Sc} are restricted to their secondary and tertiary structure. The conformational conversion of PrP^C to PrP^{Sc} is accompanied by a change in biochemical properties. While PrP^C is protease-sensitive (and is thus also named PrP^{sen} for sensitive), the pathological isoform PrP^{Sc} is more resistant to proteolytic digestion (and is thus also named PrP^{res} for resistant) (Caughey *et al.*, 1991; Horiuchi and Caughey, 1999). The controlled digestion of tissues from diseased organisms with proteinase K (PK) leads to the generation of an N-terminally truncated form of PrP^{Sc} (also named PrP27–30 according to its molecular weight on SDS-PAGE). The presence of the protease-resistant core of PrP^{Sc} is a specific and reliable molecular marker for prion disease. Recent data show that there are also protease-sensitive PrP-conformers, including oligomeric PrP-species, that do not fulfil the criteria of

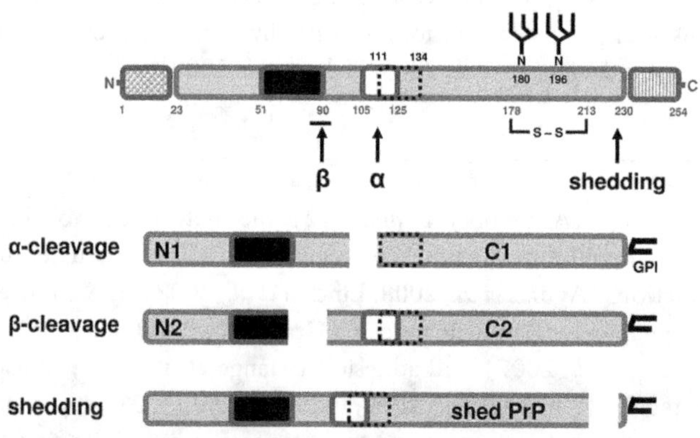

Figure 5.2 The primary sequence of murine PrP^C with important protein domains. Following removal of the N-terminal signal sequence (aa 1–22; grey dotted box) and the C-terminal signal sequence for the GPI-anchor (aa 231–254; grey striped box), mature PrP^C consists of an octameric repeat region (aa 51–90; black box), a neurotoxic domain (aa 105–125; white box), a hydrophobic core (aa 111–134; dotted box), a disulfide bridge (between aa 178 and 213), and two N-glycosylation sites (aa 180 and 196). Cleavage events are marked by arrows. (1) α-Cleavage resulting in soluble N1 (11kDa) and a membrane-bound C1 (18kDa). This cleavage occurs in the middle of the neurotoxic domain. (2) β-cleavage C-terminal of the octameric repeats leading to N2 (9kDa) and C2 (20kDa) fragments. (3) Shedding N-terminal of the GPI-anchor releases nearly full-length PrP^C into the extracellular space. Modified from Altmeppen *et al.* (2012).

PrP^{Sc} outlined above, yet associate with prion disease and contain considerable amounts of prion infectivity (Silveira *et al.*, 2005; Sandberg *et al.*, 2011; Lewis *et al.*, 2012).

Since PrP^C is anchored to the outer leaflet of membranes in neurons, it was proposed to act in signalling cascades (Mange *et al.*, 2002; Malaga-Trillo *et al.*, 2009). In fact, PrP^C plays a major role in signalling where signal transduction is either directly achieved by PrP^C or where PrP^C acts as a regulatory cofactor (Mouillet-Richard, 2000; Solforosi *et al.*, 2004; Resenberger *et al.*, 2011a; You *et al.*, 2012). In this respect it is worth to mention that oligomeric amyloid β (Aβ), the neurotoxic species involved in the pathogenesis of Alzheimer's disease, and other β-sheet-rich conformers produced in neurodegenerative diseases bind to PrP^C (Lauren *et al.*, 2009; Resenberger *et al.*, 2011a; You *et al.*, 2012). If binding of oligomeric Aβ to PrP^C, which has recently been shown to alter synaptic function by Fyn kinase activation, is the explanation to neurotoxic signalling in Alzheimer's disease is currently under debate (Lauren *et al.*, 2009; Balducci *et al.*, 2010; Calella *et al.*, 2010; Gimbel *et al.*, 2010; Kessels *et al.*, 2010; Um *et al.*, 2012).

Several other physiological functions were proposed for PrP^C. These include development and maintenance of neurons, specifically synapses (Graner *et al.*, 2000; Steele *et al.*, 2006; Hajj *et al.*, 2007), cell adhesion (Mange *et al.*, 2002; Malaga-Trillo *et al.*, 2009), neuroprotection (Chiarini *et al.*, 2002; Rambold *et al.*, 2008), and the maintenance of copper-homeostasis (Brown, 2003). The fact that PrP^C is developmentally regulated during mouse embryogenesis emphasizes its potential functions in development. As mentioned above, PrP^C does not span the membrane thus all physiological functions of PrP^C seem to require the presence of accessory molecules.

Studies investigating proteins that bind PrP^C have provided important insights into putative functions of PrP^C. In fact, the important link of PrP^C to Alzheimer's disease via binding of oligomeric Aβ to PrP^C was identified by using binder-screens with PrP^C (Lauren *et al.*, 2009). Other proteins binding to PrP^C include the amyloid precursor-like protein 1 (APLP1) a member of the amyloid precursor protein family, the antiapoptotic protein bcl-2, the 37 kDa human laminin receptor precursor, the low-density lipoprotein receptor-related protein 1, and the vesicle trafficking protein Rab7a (Jen *et al.*, 2010). Although specific modes of action upon binding of these proteins have been suggested, the physiological relevance of these interactions is not fully understood. A recently published study shows that the octarepeat domain of PrP^C binds to Argonaute proteins, essential components of microRNA (miRNA)-induced silencing complexes (miRISCs). Thus, this study links PrP^C to RNA silencing and thus implies a function of PrP^C in gene repression (Gibbings *et al.*, 2012).

The gene encoding the prion protein

The gene encoding PrP^C is termed *Prnp* in mouse and *PRNP* in humans. In humans it is located on the short arm of chromosome 20 and in mice on chromosome 2 and is encoded within a single open reading frame. The expression of PrP^C is tightly controlled by transcription factors and it was shown that specific protein 1 (Sp1) plays a role in regulating the activity of the *Prnp* promoter (Sakudo *et al.*, 2010). In human and animal genomes, *PRNP* or *Prnp* is present as a single copy gene of approximately 16 kb size.

Several partial homologues of *Prnp* have been identified. These include a protein, named Doppel (Dpl), with significant biochemical and structural homology (Behrens and Aguzzi,

2002). Dpl does not influence pathogenesis of prion diseases but rather regulates sperm generation and male reproduction. A further partial homologue of *Prnp* is Shadoo (Sho) encoded by the *SPRN* gene (Daude and Westaway, 2011). This glycoprotein is also mainly expressed in the central nervous system and has similarity to the natively unstructured N-terminus of PrP^C. During prion disease there may be a cross-regulation between Sho and PrP^{Sc} and a polymorphism in the signal peptide of SPRN constitutes a risk for sporadic CJD (Daude and Westaway, 2011).

The mRNA of *Prnp* is constitutively expressed in adult organisms and tightly regulated during development and mouse embryogenesis. As described above, mutations in host-encoded *PRNP* co-segregate with human prion diseases. The influence of polymorphisms in *PRNP* has been studied in detail and it is obvious that polymorphic codon 129 within *PRNP*, encoding for either methionine or valine, influences incubation time until onset of prion disease following prion-exposure and clinical features of the disease. An important pillar supporting the prion principle came from studies in mice lacking PrP^C, which do not develop prion disease when challenged with infectious prions (Büeler *et al.*, 1993).

Misfolding of the prion protein

Generation and tissue deposition of an abnormal isoform of host encoded PrP^C (PrP^{Sc}) is the hallmark of prion disease. Since PrP^C and PrP^{Sc} do not differ in their amino acid sequence and harbour identical post-translational modifications their differences in conformation may help to unravel mechanism of misfolding. For the globular C-terminal part of PrP^C, detailed structural information is available through the help of nuclear magnetic resonance (NMR) spectroscopy and X-ray crystallography. The amyloidogenic nature of PrP^{Sc} has been investigated using Congo Red and Thioflavin T staining, which clearly demonstrate characteristic features of PrP^{Sc} aggregates. More detailed structural data for PrP^{Sc} exist only with low-resolution since PrP^{Sc} is insoluble and cannot be purified in sufficient quantity and quality. Using far ultraviolet CD-spectra and X-ray diffraction studies of purified amyloid fibres, it was shown that the protofilament cores of PrP^{Sc} consist of 'cross-β' scaffolds where β-sheets stand parallel to the fibre-axis and β-strands are perpendicularly oriented. By using short oligopeptides of amyloid-forming proteins, crystal structure analysis was possible (Surewicz and Apostol, 2011). This revealed that stretches of four to seven amino acid residues suffice for fibril formation. This implies that amyloid-like PrP^{Sc} aggregates may grow by interdigitating β-sheets in a process resembling a zipper. Through the usage of cryoelectron microscopy and modelling approaches with protein structures common to amyloid forming proteins it has been postulated that β-sheets of PrP^{Sc} are located between amino acid 81–95 and amino acid 171, whereas the C-terminus remains preserved. It was proposed that β-sheets form a left-handed beta-helix with three PrP^{Sc} molecules forming the primary unit which constitutes the basis for scrapie-associated fibrils (Govaerts *et al.*, 2004).

At least two models have been postulated to explain the mechanism by which PrP^{Sc} induces the misfolding of PrP^C (Weissmann, 2004). In the template assisted model, the conformational change is controlled kinetically. This means that a high activation energy barrier prevents considerable spontaneous conversion. In this model, exogenously introduced PrP^{Sc} or spontaneously formed PrP^{Sc} leads to conformational change of PrP^C to yield PrP^{Sc}. This reaction, which is not described at the molecular level, involves subsequent steps of unfolding and refolding. In order to overcome the postulated high energy barrier, this process is likely to depend on an enzyme or chaperone. In the second model, which is generally known as the nucleation-polymerization or 'seeding' model, PrP^{C} and PrP^{Sc} are in equilibrium, with the amount of PrP^{C} greatly surpassing the amount of PrP^{Sc}. Only when a critical seed size is surpassed can additional PrP^{C} be recruited to the seed which can then grow in a crystal-like fashion by a nucleation-dependent polymerization process. Several studies using cell-free conversion systems support this model. Striking similarities between infectious prion disorders and other protein misfolding diseases such as Alzheimer's disease, Parkinson's disease, amyotrophic lateral sclerosis, as well as diffuse frontotemporal dementias, where protein misfolding spreads via seeding processes, imply that this process is not limited to prion disease (Prusiner, 2012).

Mechanisms of neuronal death in prion diseases

PrP^{Sc} is found in the brain in most prion diseases. However, in some genetic forms of CJD, in protease-sensitive prionopathies of humans, and in some experimental and naturally occurring prion diseases of animals, PrP^{Sc} is hardly detectable (Cronier *et al.*, 2008; Gambetti *et al.*, 2008). Thus it is under discussion if PrP^{Sc} itself or other PK-sensitive PrP-species may represent the prime neurotoxic entity. This fits to data showing an uncoupling of PrP^{Sc} amounts and titres of prion infectivity (Glatzel and Aguzzi, 2000; Barron *et al.*, 2007; Piccardo *et al.*, 2007; Sandberg *et al.*, 2011). There is an ongoing discussion on the nature of the neurotoxic species, with PK-sensitive oligomeric PrP species, transmembrane forms of the prion protein and cytosolic PrP-species as the main candidates (Sandberg *et al.*, 2011). PrP^{C} differs from other GPI-anchored proteins due to the fact that it can give rise to several different topological forms. Besides the regular form, with the polypeptide chain being fully translocated into the ER lumen and its C-terminus being attached to the lipid bilayer via the GPI-anchor, PrP^{C} can adopt transmembrane orientations. This occurs when the hydrophobic core (HC) of PrP^{C}, which is located centrally within the protein, acts as a transmembrane region (Hegde *et al.*, 1998; Stewart *et al.*, 2005). In these instances, either the N-terminus (NTMPrP) or the C-terminus (CTMPrP) are exposed towards the luminal side. CTMPrP seems to be more abundant and has neurotoxic properties (Stewart *et al.*, 2005). Cytosolic PrP-species are generated when PrP^{C} is retrotranslocated to the cytosol and it has been suggested that these forms of the prion protein may cause neurodegeneration, possibly by overwhelming the protein-clearance capacity of neurons (Ma *et al.*, 2002). Indeed, a noteworthy proportion of misfolded PrP^{C} can be found in the cytosol in transgenic mice expressing PrP^{C} lacking N-terminal secretory signal peptide and C-terminal signal sequence for addition of the GPI-anchor, yet it is unclear if this is relevant in naturally occurring prion disease (Roucou *et al.*, 2003). As for Aβ and AD, PrP^{C} itself seems to be involved in transducing neurotoxic signalling in prion disease (Cronier *et al.*, 2012). The details of neurotoxic signalling are not fully understood. Similar to AD, Fyn kinase activation or N-methyl-D-aspartate (NMDA) receptor-mediated excessive excitatory activity are being discussed in this context (Resenberger *et al.*, 2011b).

The molecular mechanism underlying neurodegeneration in prion disease are far from being fully understood, yet characterization of various potentially neurotoxic PrP-species will help to understand the execution of neurodegenerative pathways. The data discussed above do have practical implications. The fact that PrP^{Sc} amounts alone can not serve as a marker for tissue infectivity suggests that it may be wise to adapt current protocols of prion detection in tissues, since they are so far largely based on the detection of bona fide PrP^{Sc} (Safar *et al.*, 2005).

Shedding and anchorless PrP

PrP^C undergoes post-translational proteolytic processing and these cleavage events likely regulate the physiological functions of PrP^C and influence the pathophysiology of prion disease (Altmeppen *et al.*, 2011, 2012).

Proteolytic processing of PrP^C includes its α-cleavage, β-cleavage, and shedding. α-Cleavage leads to the generation of a soluble N1 fragment of 11 kDa and a membrane-bound C1 fragment of 18 kDa whereas β-cleavage produces N2 (9 kDa) and C2 (20 kDa) fragments. PrP^C-fragments produced by α-cleavage have roles related to the physiological function of PrP^C, namely neuroprotection for N1 and maintenance of myelin-integrity for C1 (Guillot-Sestier *et al.*, 2009, 2012; Bremer *et al.*, 2010; Altmeppen *et al.*, 2012). β-cleavage does not play a major role under physiological conditions, but is increased in situations of oxidative stress, i.e. in prion disease (Chen *et al.*, 1995; Jimenez-Huete *et al.*, 1998). The exact functions of PrP^C-fragments produced by β-cleavage are not entirely understood but may relate to oxidative stress (Jimenez-Huete *et al.*, 1998; McMahon *et al.*, 2001; Mange *et al.*, 2004).

Shedding of PrP^C occurs in close proximity to its GPI-anchor and results in the release of PrP^C lacking only a few C-terminal amino acids (Borchelt *et al.*, 1993). Shed and soluble forms of PrP^C are found in cell culture but also under *in vivo* conditions, i.e. in human CSF (Tagliavini *et al.*, 1992) and blood (Perini *et al.*, 1996; MacGregor *et al.*, 1999; Parizek *et al.*, 2001). A-disintegrin-and-metalloproteinase (ADAM) 10 is the main sheddase *in vitro* and *in vivo* (Taylor *et al.*, 2009; Altmeppen *et al.*, 2011).

PrP^C shedding may influence prion pathogenesis in two ways thus having a dual role in the course of disease. On the one hand, it regulates PrP^C levels at the plasma membrane where it reduces the substrate for conversion (Marella *et al.*, 2002; Heiseke *et al.*, 2008). Indeed, increased release of surface PrP^C by treatment with phospholipase C or with the drug filipin interferes with the formation of PrP^{Sc} (Caughey and Raymond, 1991; Borchelt *et al.*, 1992; Enari *et al.*, 2001; Marella *et al.*, 2002). In line with this, mice that overexpress ADAM10 have prolonged incubation times when prion infected (Endres *et al.*, 2009).

On the other hand, shedding of PrP^C or PrP^{Sc} may favour prion disease by facilitating prion spread throughout the nervous system. In line with this, ADAM10 can shed PrP^{Sc} in infected neuroblastoma cells and anchorless mutant versions of PrP^C can be converted to PrP^{Sc} in cell culture and cell-free systems (Rogers *et al.*, 1993; Kocisko *et al.*, 1994; Taylor *et al.*, 2009). These cell culture-based data are supported by results showing that transgenic expression of anchorless, secreted PrP leads to enhanced deposition of PrP^{Sc} when mice are prion infected (Chesebro *et al.*, 2005). However, since incubation times to terminal prion disease in these 'anchorless PrP' mice was prolonged, neurodegeneration seems to depend on membrane-anchoring of PrP^C (Chesebro *et al.*, 2010), a finding that nicely fits to the involvement of different entities responsible for neurotoxicity on the one and prion propagation on the other hand (see above under 'Mechanisms of neuronal death in prion diseases') (Sandberg *et al.*, 2011).

Prions and glia

The prime target of prion diseases is the central nervous system and usually neuronal damage and neuronal loss is profound (Schoch *et al.*, 2005). Yet in some prion diseases, neuronal loss is minimal. Astrogliosis and microgliosis on the other hand are consistent findings in all prion diseases. In line with this, recent data point out the importance of astrocytic involvement in prion disease. Astrocytes are capable of supporting prion replication and astrocytic reaction influences the neuropathological and clinical presentation of prion disease (Raeber *et al.*, 1997; Krasemann *et al.*, 2012). Another cell type potentially implicated in execution of neurotoxicity in prion diseases is the microglial cell (Marella and Chabry, 2004). These cells can be viewed as the brain's immunocompetent macrophages and are implicated in a wide range of CNS diseases including multiple sclerosis and Alzheimer's disease (Streit, 2010). Cell-culture based experiments suggest that the activation status of microglia is critical for transducing neurotoxic effects in prion disease (Bate *et al.*, 2005). In order to fully understand pathways of neuronal cell death, it will therefore be essential to investigate the individual contributions of neurons, astrocytes and microglia.

Prion diseases and non-central nervous system organs

When assessing the distribution of pathological lesions, the amounts of PrP^{Sc} or prion infectivity and expression of PrP^{C} it becomes obvious that the central nervous system stands apart from other organ systems (Schoch *et al.*, 2006). Yet, this should not hide the fact that organ systems besides the brain play important roles in prion disease. In fact, most acquired forms of prion diseases are transmitted through peripheral uptake of the prions and reservoirs of PrP^{Sc} or prion infectivity exist in lymphoid organs and the muscle (Glatzel and Aguzzi, 2001).

Our knowledge on the pathways that prions take in order to invade the brain and our understanding of what it takes to establish and maintain prion-replication in peripheral sites is profound (Glatzel and Aguzzi, 2001).

Regarding neuroinvasion of prions, it is clear that this process involves several distinct phases. In the first phase, colonization of the lymphoreticular system occurs. Here, follicular dendritic cells within lymphoid follicles play a central role in capturing and replicating prions. In the second phase, prions spread through the peripheral nervous system to invade the brain. Depending on the route of prion administration, both the sympathetic and the parasympathetic nervous systems are implied in this process (Glatzel *et al.*, 2001).

Colonization of the neuromuscular compartment by prions is of interest due to the fact that this represents the prime organ system entering the human food chain in the case of farmed animals. In most prion diseases of humans and animals, the presence of PrP^{Sc} or prion infectivity can be found in the neuromuscular compartment (Glatzel *et al.*, 2003; Krasemann *et al.*, 2010). Yet it is reassuring that amounts of PrP^{Sc} or prion infectivity are low and that prion colonization occurs late in the course of disease (Balkema-Buschmann *et al.*, 2011).

References

Aguzzi, A., Baumann, F., and Bremer, J. (2008). The prion's elusive reason for being. Annu. Rev. Neurosci. *31*, 439–477.

Altmeppen, H.C., Prox, J., Puig, B., Kluth, M.A., Bernreuther, C., Thurm, D., Jorissen, E., Petrowitz, B., Bartsch, U., De Strooper, B., *et al.* (2011). Lack of a-disintegrin-and-metalloproteinase ADAM10 leads to intracellular accumulation and loss of shedding of the cellular prion protein *in vivo.* Mol. Neurodegener. *6,* 36.

Altmeppen, H.C., Puig, B., Dohler, F., Thurm, D.K., Falker, C., Krasemann, S., and Glatzel, M. (2012). Proteolytic processing of the prion protein in health and disease. Am. J. Neurodegener. Dis. *1,* 15–31.

Balducci, C., Beeg, M., Stravalaci, M., Bastone, A., Sclip, A., Biasini, E., Tapella, L., Colombo, L., Manzoni, C., Borsello, T., *et al.* (2010). Synthetic amyloid-beta oligomers impair long-term memory independently of cellular prion protein. Proc. Natl. Acad. Sci. U.S.A. *107,* 2295–2300.

Balkema-Buschmann, A., Eiden, M., Hoffmann, C., Kaatz, M., Ziegler, U., Keller, M., and Groschup, M.H. (2011). BSE infectivity in the absence of detectable PrP(Sc) accumulation in the tongue and nasal mucosa of terminally diseased cattle. J. Gen. Virol. *92,* 467–476.

Barron, R.M., Campbell, S.L., King, D., Bellon, A., Chapman, K.E., Williamson, R.A., and Manson, J.C. (2007). High titers of transmissible spongiform encephalopathy infectivity associated with extremely low levels of PrPSc *in vivo.* J. Biol. Chem. *282,* 35878–35886.

Bate, C., Boshuizen, R., and Williams, A. (2005). Microglial cells kill prion-damaged neurons *in vitro* by a CD14-dependent process. J. Neuroimmunol. *170,* 62–70.

Behrens, A., and Aguzzi, A. (2002). Small is not beautiful: antagonizing functions for the prion protein PrP(C) and its homologue Dpl. Trends Neurosci. *25,* 150–154.

Biasini, E., Turnbaugh, J.A., Unterberger, U., and Harris, D.A. (2012). Prion protein at the crossroads of physiology and disease. Trends Neurosci. *35,* 92–103.

Borchelt, D.R., Taraboulos, A., and Prusiner, S.B. (1992). Evidence for synthesis of scrapie prion proteins in the endocytic pathway. J. Biol. Chem. *267,* 16188–16199.

Borchelt, D.R., Rogers, M., Stahl, N., Telling, G., and Prusiner, S.B. (1993). Release of the cellular prion protein from cultured cells after loss of its glycoinositol phospholipid anchor. Glycobiology *3,* 319–329.

Bremer, J., Baumann, F., Tiberi, C., Wessig, C., Fischer, H., Schwarz, P., Steele, A.D., Toyka, K.V., Nave, K.A., Weis, J., *et al.* (2010). Axonal prion protein is required for peripheral myelin maintenance. Nat. Neurosci. *13,* 310–318.

Brown, D.R. (2003). Prion protein expression modulates neuronal copper content. J. Neurochem. *87,* 377–385.

Brown, P., Preece, M., Brandel, J.P., Sato, T., McShane, L., Zerr, I., Fletcher, A., Will, R.G., Pocchiari, M., Cashman, N.R., *et al.* (2000). Iatrogenic Creutzfeldt–Jakob disease at the millennium. Neurology *55,* 1075–1081.

Brugger, B., Graham, C., Leibrecht, I., Mombelli, E., Jen, A., Wieland, F., and Morris, R. (2004). The membrane domains occupied by glycosylphosphatidylinositol-anchored prion protein and Thy-1 differ in lipid composition. J. Biol. Chem. *279,* 7530–7536.

Büeler, H.R., Aguzzi, A., Sailer, A., Greiner, R.A., Autenried, P., Aguet, M., and Weissmann, C. (1993). Mice devoid of PrP are resistant to scrapie. Cell *73,* 1339–1347.

Calella, A.M., Farinelli, M., Nuvolone, M., Mirante, O., Moos, R., Falsig, J., Mansuy, I.M., and Aguzzi, A. (2010). Prion protein and Abeta-related synaptic toxicity impairment. EMBO Mol. Med. *2,* 306–314.

Capobianco, R., Casalone, C., Suardi, S., Mangieri, M., Miccolo, C., Limido, L., Catania, M., Rossi, G., Di Fede, G., Giaccone, G., *et al.* (2007). Conversion of the BASE prion strain into the BSE strain: the origin of BSE? PLoS Pathog. *3,* e31.

Caughey, B., and Raymond, G.J. (1991). The scrapie-associated form of PrP is made from a cell surface precursor that is both protease- and phospholipase-sensitive. J. Biol. Chem. *266,* 18217–18223.

Caughey, B., Raymond, G.J., Ernst, D., and Race, R.E. (1991). N-terminal truncation of the scrapie-associated form of PrP by lysosomal protease(s): implications regarding the site of conversion of PrP to the protease-resistant state. J. Virol. *65,* 6597–6603.

Chen, S.G., Teplow, D.B., Parchi, P., Teller, J.K., Gambetti, P., and Autilio-Gambetti, L. (1995). Truncated forms of the human prion protein in normal brain and in prion diseases. J. Biol. Chem. *270,* 19173–19180.

Chesebro, B., Trifilo, M., Race, R., Meade-White, K., Teng, C., LaCasse, R., Raymond, L., Favara, C., Baron, G., Priola, S., *et al.* (2005). Anchorless prion protein results in infectious amyloid disease without clinical scrapie. Science *308,* 1435–1439.

Chesebro, B., Race, B., Meade-White, K., Lacasse, R., Race, R., Klingeborn, M., Striebel, J., Dorward, D., McGovern, G., and Jeffrey, M. (2010). Fatal transmissible amyloid encephalopathy: a new type of prion disease associated with lack of prion protein membrane anchoring. PLoS Pathog. *6,* e1000800.

Chiarini, L.B., Freitas, A.R., Zanata, S.M., Brentani, R.R., Martins, V.R., and Linden, R. (2002). Cellular prion protein transduces neuroprotective signals. EMBO J. *21,* 3317–3326.

Colby, D.W., and Prusiner, S.B. (2011). Prions. Cold Spring Harb. Perspect. Biol. *3*, a006833.

Collinge, J., Whitfield, J., McKintosh, E., Beck, J., Mead, S., Thomas, D.J., and Alpers, M.P. (2006). Kuru in the 21st century – an acquired human prion disease with very long incubation periods. Lancet *367*, 2068–2074.

Cronier, S., Gros, N., Tattum, M.H., Jackson, G.S., Clarke, A.R., Collinge, J., and Wadsworth, J.D. (2008). Detection and characterization of proteinase K-sensitive disease-related prion protein with thermolysin. Biochem. J. *416*, 297–305.

Cronier, S., Carimalo, J., Schaeffer, B., Jaumain, E., Beringue, V., Miquel, M.C., Laude, H., and Peyrin, J.M. (2012). Endogenous prion protein conversion is required for prion-induced neuritic alterations and neuronal death. FASEB J. *26*, 3854–3861.

Daude, N., and Westaway, D. (2011). Biological properties of the PrP-like Shadoo protein. Front. Biosci. *16*, 1505–1516.

Enari, M., Flechsig, E., and Weissmann, C. (2001). Scrapie prion protein accumulation by scrapie-infected neuroblastoma cells abrogated by exposure to a prion protein antibody. Proc. Natl. Acad. Sci. U.S.A. *98*, 9295–9299.

Endres, K., Mitteregger, G., Kojro, E., Kretzschmar, H., and Fahrenholz, F. (2009). Influence of ADAM10 on prion protein processing and scrapie infectiosity *in vivo*. Neurobiol. Dis. *36*, 233–241.

Gajdusek, D.C., and Reid, L.H. (1961). Studies on kuru. IV. The kuru pattern in Moke, a representative Fore village. Am. J. Trop. Med. Hyg. *10*, 628–638.

Gambetti, P., Dong, Z., Yuan, J., Xiao, X., Zheng, M., Alshekhlee, A., Castellani, R., Cohen, M., Barria, M.A., Gonzalez-Romero, D., *et al.* (2008). A novel human disease with abnormal prion protein sensitive to protease. Ann. Neurol. *63*, 697–708.

Geissen, M., Krasemann, S., Matschke, J., and Glatzel, M. (2007). Understanding the natural variability of prion diseases. Vaccine *25*, 5631–5636.

Gibbings, D., Leblanc, P., Jay, F., Pontier, D., Michel, F., Schwab, Y., Alais, S., Lagrange, T., and Voinnet, O. (2012). Human prion protein binds Argonaute and promotes accumulation of microRNA effector complexes. Nat. Struct. Mol. Biol. *19*, 517–524, S511.

Gimbel, D.A., Nygaard, H.B., Coffey, E.E., Gunther, E.C., Lauren, J., Gimbel, Z.A., and Strittmatter, S.M. (2010). Memory impairment in transgenic Alzheimer mice requires cellular prion protein. J. Neurosci. *30*, 6367–6374.

Glatzel, M., and Aguzzi, A. (2000). PrP(C) expression in the peripheral nervous system is a determinant of prion neuroinvasion. J. Gen. Virol. *81*, 2813–2821.

Glatzel, M., and Aguzzi, A. (2001). The shifting biology of prions. Brain Res. Brain Res. Rev. *36*, 241–248.

Glatzel, M., Heppner, F.L., Albers, K.M., and Aguzzi, A. (2001). Sympathetic innervation of lymphoreticular organs is rate limiting for prion neuroinvasion. Neuron *31*, 25–34.

Glatzel, M., Rogivue, C., Ghani, A., Streffer, J.R., Amsler, L., and Aguzzi, A. (2002). Incidence of Creutzfeldt–Jakob disease in Switzerland. Lancet *360*, 139–141.

Glatzel, M., Abela, E., Maissen, M., and Aguzzi, A. (2003). Extraneural pathologic prion protein in sporadic Creutzfeldt–Jakob disease. N. Engl. J. Med. *349*, 1812–1820.

Govaerts, C., Wille, H., Prusiner, S.B., and Cohen, F.E. (2004). Evidence for assembly of prions with left-handed beta-helices into trimers. Proc. Natl. Acad. Sci. U.S.A. *101*, 8342–8347.

Graner, E., Mercadante, A.F., Zanata, S.M., Forlenza, O.V., Cabral, A.L., Veiga, S.S., Juliano, M.A., Roesler, R., Walz, R., Minetti, A., *et al.* (2000). Cellular prion protein binds laminin and mediates neuritogenesis. Brain Res. Mol. Brain Res. *76*, 85–92.

Hajj, G.N., Lopes, M.H., Mercadante, A.F., Veiga, S.S., da Silveira, R.B., Santos, T.G., Ribeiro, K.C., Juliano, M.A., Jacchieri, S.G., Zanata, S.M., *et al.* (2007). Cellular prion protein interaction with vitronectin supports axonal growth and is compensated by integrins. J. Cell Sci. *120*, 1915–1926.

Haley, N.J., Seelig, D.M., Zabel, M.D., Telling, G.C., and Hoover, E.A. (2009). Detection of CWD prions in urine and saliva of deer by transgenic mouse bioassay. PLoS One *4*, e4848.

Hegde, R.S., Mastrianni, J.A., Scott, M.R., DeFea, K.A., Tremblay, P., Torchia, M., DeArmond, S.J., Prusiner, S.B., and Lingappa, V.R. (1998). A transmembrane form of the prion protein in neurodegenerative disease. Science *279*, 827–834.

Heiseke, A., Schobel, S., Lichtenthaler, S.F., Vorberg, I., Groschup, M.H., Kretzschmar, H., Schatzl, H.M., and Nunziante, M. (2008). The novel sorting nexin SNX33 interferes with cellular PrP formation by modulation of PrP shedding. Traffic *9*, 1116–1129.

Hill, A.F., Desbruslais, M., Joiner, S., Sidle, K.C., Gowland, I., Collinge, J., Doey, L.J., and Lantos, P. (1997). The same prion strain causes vCJD and BSE. Nature [letter] *389*, 448–450.

Horiuchi, M., and Caughey, B. (1999). Specific binding of normal prion protein to the scrapie form via a localized domain initiates its conversion to the protease-resistant state. EMBO J. *18*, 3193–3203.

Ironside, J.W. (1998). Prion diseases in man. J. Pathol. *186*, 227–234.

Jakob, A. (1921). Über eigenartige Erkrankungen des Zentralnervensystems mit bemerkenswertem anatomischem Befunde. (Spastische Pseudosklerose-Encephalomyelopathie mit disseminierten Degenerationsherden). Z ges Neurol. Psychiatr. *64*, 147–228.

Jen, A., Parkyn, C.J., Mootoosamy, R.C., Ford, M.J., Warley, A., Liu, Q., Bu, G., Baskakov, I.V., Moestrup, S., McGuinness, L., *et al.* (2010). Neuronal low-density lipoprotein receptor-related protein 1 binds and endocytoses prion fibrils via receptor cluster 4. J. Cell Sci. *123*, 246–255.

Jimenez-Huete, A., Lievens, P.M., Vidal, R., Piccardo, P., Ghetti, B., Tagliavini, F., Frangione, B., and Prelli, F. (1998). Endogenous proteolytic cleavage of normal and disease-associated isoforms of the human prion protein in neural and non-neural tissues. Am. J. Pathol. *153*, 1561–1572.

Kessels, H.W., Nguyen, L.N., Nabavi, S., and Malinow, R. (2010). The prion protein as a receptor for amyloid-beta. Nature *466*, E3–4; discussion E4–5.

Kocisko, D.A., Come, J.H., Priola, S.A., Chesebro, B., Raymond, G.J., Lansbury, P.T., and Caughey, B. (1994). Cell-free formation of protease-resistant prion protein. Nature *370*, 471–474.

Krasemann, S., Neumann, M., Geissen, M., Bodemer, W., Kaup, F.J., Schulz-Schaeffer, W., Morel, N., Aguzzi, A., and Glatzel, M. (2010). Preclinical deposition of pathological prion protein in muscle of experimentally infected primates. PLoS ONE *5*, e13906.

Krasemann, S., Neumann, M., Luepke, J.P., Grashorn, J., Wurr, S., Stocking, C., and Glatzel, M. (2012). Persistent retroviral infection with MoMuLV influences neuropathological signature and phenotype of prion disease. Acta Neuropathol. *124*, 111–126.

Lauren, J., Gimbel, D.A., Nygaard, H.B., Gilbert, J.W., and Strittmatter, S.M. (2009). Cellular prion protein mediates impairment of synaptic plasticity by amyloid-beta oligomers. Nature *457*, 1128–1132.

Lewis, V., Haigh, C.L., Masters, C.L., Hill, A.F., Lawson, V.A., and Collins, S.J. (2012). Prion subcellular fractionation reveals infectivity spectrum, with a high titre-low PrPres level disparity. Mol. Neurodegener. *7*, 18.

Linden, R., Martins, V.R., Prado, M.A., Cammarota, M., Izquierdo, I., and Brentani, R.R. (2008). Physiology of the prion protein. Physiol. Rev. *88*, 673–728.

Llewelyn, C.A., Hewitt, P.E., Knight, R.S., Amar, K., Cousens, S., Mackenzie, J., and Will, R.G. (2004). Possible transmission of variant Creutzfeldt–Jakob disease by blood transfusion. Lancet *363*, 417–421.

Ma, J., Wollmann, R., and Lindquist, S. (2002). Neurotoxicity and neurodegeneration when PrP accumulates in the cytosol. Science *298*, 1781–1785.

MacGregor, I., Hope, J., Barnard, G., Kirby, L., Drummond, O., Pepper, D., Hornsey, V., Barclay, R., Bessos, H., Turner, M., *et al.* (1999). Application of a time-resolved fluoroimmunoassay for the analysis of normal prion protein in human blood and its components. Vox Sang *77*, 88–96.

McMahon, H.E., Mange, A., Nishida, N., Creminon, C., Casanova, D., and Lehmann, S. (2001). Cleavage of the amino terminus of the prion protein by reactive oxygen species. J. Biol. Chem. *276*, 2286–2291.

Malaga-Trillo, E., Solis, G.P., Schrock, Y., Geiss, C., Luncz, L., Thomanetz, V., and Stuermer, C.A. (2009). Regulation of embryonic cell adhesion by the prion protein. PLoS Biol. *7*, e55.

Mange, A., Milhavet, O., Umlauf, D., Harris, D., and Lehmann, S. (2002). PrP-dependent cell adhesion in N2a neuroblastoma cells. FEBS Lett. *514*, 159–162.

Mange, A., Beranger, F., Peoc'h, K., Onodera, T., Frobert, Y., and Lehmann, S. (2004). Alpha- and beta-cleavages of the amino-terminus of the cellular prion protein. Biol. Cell *96*, 125–132.

Marella, M., and Chabry, J. (2004). Neurons and astrocytes respond to prion infection by inducing microglia recruitment. J. Neurosci. *24*, 620–627.

Marella, M., Lehmann, S., Grassi, J., and Chabry, J. (2002). Filipin prevents pathological prion protein accumulation by reducing endocytosis and inducing cellular PrP release. J. Biol. Chem. *277*, 25457–25464.

Mouillet-Richard, S., Schneider, B., Pradines, E., Pietri, M., Ermonval, M., Grassi, J., Richards, J.G., Mutel, V., Launay, J.M., and Kellermann, O. (2007). Cellular prion protein signaling in serotonergic neuronal cells. Ann. N.Y. Acad. Sci. *1096*, 106–119.

Parizek, P., Roeckl, C., Weber, J., Flechsig, E., Aguzzi, A., and Raeber, A.J. (2001). Similar turnover and shedding of the cellular prion protein in primary lymphoid and neuronal cells. J. Biol. Chem. *276*, 44627–44632.

Parry, H. (1960). Scrapie: A transmissible hereditary disease of sheep. Nature *185*, 441–443.

Perini, F., Vidal, R., Ghetti, B., Tagliavini, F., Frangione, B., and Prelli, F. (1996). PrP27–30 is a normal soluble prion protein fragment released by human platelets. Biochem. Biophys. Res. Commun. *223*, 572–577.

Piccardo, P., Manson, J.C., King, D., Ghetti, B., and Barron, R.M. (2007). Accumulation of prion protein in the brain that is not associated with transmissible disease. Proc. Natl. Acad. Sci. U.S.A. *104*, 4712–4717.

Prusiner, S.B. (2012). Cell biology. A unifying role for prions in neurodegenerative diseases. Science *336*, 1511–1513.

Puig, B., Altmeppen, H.C., Thurm, D., Geissen, M., Conrad, C., Braulke, T., and Glatzel, M. (2011). N-Glycans and glycosylphosphatidylinositol-anchor act on polarized sorting of mouse PrP in Madin–Darby canine kidney cells. PLoS ONE *6*, e24624.

Raeber, A.J., Race, R.E., Brandner, S., Priola, S.A., Sailer, A., Bessen, R.A., Mucke, L., Manson, J., Aguzzi, A., Oldstone, M.B., *et al.* (1997). Astrocyte-specific expression of hamster prion protein (PrP) renders PrP knockout mice susceptible to hamster scrapie. EMBO J. *16*, 6057–6065.

Rambold, A.S., Muller, V., Ron, U., Ben-Tal, N., Winklhofer, K.F., and Tatzelt, J. (2008). Stress-protective signalling of prion protein is corrupted by scrapie prions. EMBO J. *27*, 1974–1984.

Resenberger, U.K., Harmeier, A., Woerner, A.C., Goodman, J.L., Muller, V., Krishnan, R., Vabulas, R.M., Kretzschmar, H.A., Lindquist, S., Hartl, F.U., *et al.* (2011a). The cellular prion protein mediates neurotoxic signalling of beta-sheet-rich conformers independent of prion replication. EMBO J. *30*, 2057–2070.

Resenberger, U.K., Winklhofer, K.F., and Tatzelt, J. (2011b). Neuroprotective and neurotoxic signaling by the prion protein. Top. Curr. Chem. *305*, 101–119.

Riek, R., Hornemann, S., Wider, G., Glockshuber, R., and Wüthrich, K. (1997). NMR characterization of the full-length recombinant murine prion protein, mPrP(23–231). FEBS Lett. *413*, 282–288.

Rogers, M., Yehiely, F., Scott, M., and Prusiner, S.B. (1993). Conversion of truncated and elongated prion proteins into the scrapie isoform in cultured cells. Proc. Natl. Acad. Sci. U.S.A. *90*, 3182–3186.

Roucou, X., Guo, Q., Zhang, Y., Goodyer, C.G., and LeBlanc, A.C. (2003). Cytosolic prion protein is not toxic and protects against Bax-mediated cell death in human primary neurons. J. Biol. Chem. *278*, 40877–40881.

Safar, J.G., Geschwind, M.D., Deering, C., Didorenko, S., Sattavat, M., Sanchez, H., Serban, A., Vey, M., Baron, H., Giles, K., *et al.* (2005). Diagnosis of human prion disease. Proc. Natl. Acad. Sci. U.S.A. *102*, 3501–3506.

Sakudo, A., Xue, G., Kawashita, N., Ano, Y., Takagi, T., Shintani, H., Tanaka, Y., Onodera, T., and Ikuta, K. (2010). Structure of the prion protein and its gene: an analysis using bioinformatics and computer simulation. Curr. Protein Pept. Sci. *11*, 166–179.

Sandberg, M.K., Al-Doujaily, H., Sharps, B., Clarke, A.R., and Collinge, J. (2011). Prion propagation and toxicity *in vivo* occur in two distinct mechanistic phases. Nature *470*, 540–542.

Schoch, G., Seeger, H., Bogousslavsky, J., Tolnay, M., Janzer, R.C., Aguzzi, A., and Glatzel, M. (2005). Analysis of prion strains by PrP(Sc) profiling in sporadic Creutzfeldt–Jakob disease. PLoS Med. *3*, e14.

Seeger, H., Heikenwalder, M., Zeller, N., Kranich, J., Schwarz, P., Gaspert, A., Seifert, B., Miele, G., and Aguzzi, A. (2005). Coincident scrapie infection and nephritis lead to urinary prion excretion. Science *310*, 324–326.

Sigurdson, C.J., Mathiason, C.K., Perrott, M.R., Eliason, G.A., Spraker, T.R., Glatzel, M., Manco, G., Bartz, J.C., Miller, M.W., and Hoover, E.A. (2008). Experimental chronic wasting disease (CWD) in the ferret. J. Comp. Pathol. *138*, 189–196.

Silveira, J.R., Raymond, G.J., Hughson, A.G., Race, R.E., Sim, V.L., Hayes, S.F., and Caughey, B. (2005). The most infectious prion protein particles. Nature *437*, 257–261.

Solforosi, L., Criado, J.R., McGavern, D.B., Wirz, S., Sanchez-Alavez, M., Sugama, S., DeGiorgio, L.A., Volpe, B.T., Wiseman, E., Abalos, G., *et al.* (2004). Cross-linking cellular prion protein triggers neuronal apoptosis *in vivo*. Science *303*, 1514–1516.

Steele, A.D., Emsley, J.G., Ozdinler, P.H., Lindquist, S., and Macklis, J.D. (2006). Prion protein (PrPc) positively regulates neural precursor proliferation during developmental and adult mammalian neurogenesis. Proc. Natl. Acad. Sci. U.S.A. *103*, 3416–3421.

Stewart, R.S., Piccardo, P., Ghetti, B., and Harris, D.A. (2005). Neurodegenerative illness in transgenic mice expressing a transmembrane form of the prion protein. J. Neurosci. *25*, 3469–3477.

Stoeck, K., Hess, K., Amsler, L., Eckert, T., Zimmermann, D., Aguzzi, A., and Glatzel, M. (2008). Heightened incidence of sporadic Creutzfeldt–Jakob disease is associated with a shift in clinicopathological profiles. J. Neurol. *255*, 1464–1472.

Streit, W.J. (2010). Microglial activation and neuroinflammation in Alzheimer's disease: a critical examination of recent history. Front. Aging Neurosci. *2*, 22.

Surewicz, W.K., and Apostol, M.I. (2011). Prion protein and its conformational conversion: a structural perspective. Top. Curr. Chem. *305*, 135–167.

Tagliavini, F., Prelli, F., Porro, M., Salmona, M., Bugiani, O., and Frangione, B. (1992). A soluble form of prion protein in human cerebrospinal fluid: implications for prion-related encephalopathies. Biochem. Biophys. Res. Commun. *184*, 1398–1404.

Taylor, D.R., and Hooper, N.M. (2006). The prion protein and lipid rafts. Mol. Membr. Biol. *23*, 89–99.

Taylor, D.R., Parkin, E.T., Cocklin, S.L., Ault, J.R., Ashcroft, A.E., Turner, A.J., and Hooper, N.M. (2009). Role of ADAMs in the ectodomain shedding and conformational conversion of the prion protein. J. Biol. Chem. *284*, 22590–22600.

Um, J.W., Nygaard, H.B., Heiss, J.K., Kostylev, M.A., Stagi, M., Vortmeyer, A., Wisniewski, T., Gunther, E.C., and Strittmatter, S.M. (2012). Alzheimer amyloid-beta oligomer bound to postsynaptic prion protein activates Fyn to impair neurons. Nat Neurosci. *15*, 1227–1235.

Weissmann, C. (2004). The state of the prion. Nat. Rev. Microbiol. *2*, 861–871.

Wells, G.A., Scott, A.C., Johnson, C.T., Gunning, R.F., Hancock, R.D., Jeffrey, M., Dawson, M., and Bradley, R. (1987). A novel progressive spongiform encephalopathy in cattle. VetRec *121*, 419–420.

Will, R.G. (2003). Acquired prion disease: iatrogenic CJD, variant CJD, kuru. Br. Med. Bull. *66*, 255–265.

Will, R.G., Ironside, J.W., Zeidler, M., Cousens, S.N., Estibeiro, K., Alperovitch, A., Poser, S., Pocchiari, M., Hofman, A., and Smith, P.G. (1996). A new variant of Creutzfeldt–Jakob disease in the UK. Lancet *347*, 921–925.

Wroe, S.J., Pal, S., Siddique, D., Hyare, H., Macfarlane, R., Joiner, S., Linehan, J.M., Brandner, S., Wadsworth, J.D., Hewitt, P., *et al.* (2006). Clinical presentation and pre-mortem diagnosis of variant Creutzfeldt–Jakob disease associated with blood transfusion: a case report. Lancet *368*, 2061–2067.

You, H., Tsutsui, S., Hameed, S., Kannanayakal, T.J., Chen, L., Xia, P., Engbers, J.D., Lipton, S.A., Stys, P.K., and Zamponi, G.W. (2012). Abeta neurotoxicity depends on interactions between copper ions, prion protein, and N-methyl-D-aspartate receptors. Proc. Natl. Acad. Sci. U.S.A. *109*, 1737–1742.

Inactivation of Prion and Endotoxins

6

Hideharu Shintani and Gerald McDonnell

Abstract

Several inactivation procedures to prion and endotoxins are reported so far. Most of these methods are not applicable to re-usable medical devices due to failure of achievement of material and functional compatibility. Gas plasma inactivation procedure for prion and endotoxin was studied and attain both sterility assurance level (SAL) of 10^{-6} and material and functional compatibility in ease.

Inactivation and degradation of prions

'About what prion is' will be discussed elsewhere in this book, so in this chapter we will discuss about prion inactivation, especially the specific procedure to attain both 10^{-6} sterility assurance level (SAL) and material and functional compatibility, which sterilization validation and good manufacturing practices (GMP) required.

As a typical example of an iatrogenic disease, prions have been transmitted via instruments used for neurosurgical procedures, from corneal implants and from pituitary growth hormone. The infectious agents are probably small protein particles, which are highly resistant to heat, ethylene oxide gas, glutaraldehyde and formaldehyde and so on (Fichet *et al.*, 2004). High concentrations of sodium hypochlorite (20,000 ppm, average Cl) for 1 hour are effective but are corrosive to most medical devices. Moist heat sterilization at 134–138°C for 18 minutes or 132°C for 1 hour is conducted. These procedures are effective against all strains and prions, but material and functional compatibility is not achieved. Other effective procedures to prions include autoclaving at 121°C for 30 minutes in the presence of 1 N sodium hydroxide (NaOH), autoclaving at 134°C for 18 minutes in the presence of 1 N NaOH, soaking in 2.0% sodium hypochlorite for 60 minutes, 1 N NaOH for 60 minutes, 4 M guanidine thiocyanate, or 0.9% phenolic (LpH) for 30 minutes. These highly corrosive procedures can cause significant damage to surgical instruments and therefore, are rarely used. Thorough cleaning is of great importance, owing to the doubts on the effectiveness of inactivation methods and the probable variation in the resistance of different prions to heat and disinfectants.

Prions are highly resistant to most physical and chemical agents, with even greater resistance than bacterial spores in some cases (Table 6.1; Taylor, 2004; McDonnell, 2007a); thus, unconventional agents are believed to be highly resistant to many chemical disinfection and physical sterilization processes, including ultraviolet and ionizing radiation and high

Table 6.1 Decreasing order of resistance of microorganisms to disinfectants and sterilants

Prions
Bacterial spores
Mycobacteria
Non-enveloped viruses
Fungi
Vegetative bacteria
Enveloped viruses

temperatures. At present little is known about the mechanisms of inactivation of or mechanisms of resistance by these unconventional agents.

It is important to note that crude preparations (brain homogenates from infected animals) have been used to investigate the efficacies of various biocides and biocidal processes against prions. The presence of extraneous materials (particularly a high concentration of lipid associated with brain tissue) could, at least to some extent, mask the true efficacies of these processes against the infectious agent. For disinfection of these crude extracts, there is currently no known decontamination procedure that can guarantee the complete absence of infectivity in prion-infected tissues. The most effective process is boiling or superheating under pressure of the tissues in concentrated solutions of sodium hydroxide (1 to 2 N), which over time can totally dissolve any proteins present. In contrast, prions can survive harsh acid treatment. Formaldehyde, unbuffered glutaraldehyde (acidic pH) and ethylene oxide gas have little effect on infectivity, although chlorine-releasing agents (especially hypochlorites), sodium hydroxide, some phenols, and guanidine thiocyanate are more effective. Lower concentrations of hydroxides (NaOH and KOH) and hypochlorides have been shown to be effective against surface prion contamination, in combination with surfactants and other formulation effects. Extended steam sterilization is effective, although hydration of the prion-infected material appears to be important for optimal inactivation of prions. Further research is required on developing formulations and processes against prions, including the use of oxidizing agents, like gaseous hydrogen peroxide. In case of gaseous hydrogen peroxide exposure, produced OH radicals attack to amino acids with benzene ring (phenylalanine, tyrosine, tryptophan), resulting in denaturation of prion protein due to failure of maintenance of high-order structure because of incorrect hydrogen bonding between amino acids.

Prions are hydrophobic proteins and have been shown to have affinity for surfaces, including metals and plastics. In most surface disinfection recommendations for use against prions, cleaning is considered a key step to remove most of the contamination prior to chemical or heat inactivation of the prions. However, certain cleaning formulations have been shown to increase the intrinsic resistance of prions or prion-contaminated materials by an unknown mechanism. This has also been observed in treatment with some biocides, including formaldehyde, presumably due to protein fixation. Although prions are considered resistant to proteases, various proteases, including proteinase K and keratinases, have been shown to degrade prions over time, depending on their concentrations and the exposure conditions.

Prions aggregate to form protein particles or fibrils within various tissues but are predominantly observed within brain tissue. These particles are hydrophobic and are associated with

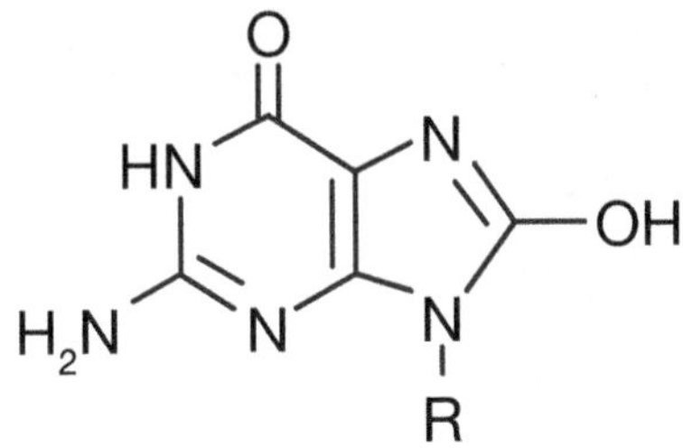

Figure 6.1 Chemical structure of 8-hydroxyguanine.

various cellular materials present within the contaminated tissue; for example, brain tissue contains a high concentration of lipid materials. These effects create a penetration challenge for the biocidal process. Various biocides and biocidal processes have been shown to be ineffective against prions. In the case of radiation sterilization, the mechanism of resistance is the absence of any nucleic acid bases associated with prion and endotoxin inactivation (Fairand, 2002). It is concretely speculated due to lack of formation of hydroxylation of DNA and RNA bases such as 8-OHG (8-hydroxyguanine; Fig. 6.1; Halliwell and Gutteridge, 2007) to prions and endotoxins, which result in failure of regular hydrogen bonding formation of nucleic acids among adenine=tymine (A=T), adenine=uracil (A=U) and guanine≡cytosine (G≡C). In addition, radiation sterilization can damage other macromolecules, including proteins due to cleavage of hydrogen bonding to form higher order structures. Biocides that have a cross-linking or fixing mode of action, including alcohols and aldehydes, have also been shown to be ineffective against prions, presumably due to a lack of any degradative effect on the target protein but actual cross-linking with other associated extraneous materials. Biocides or biocidal processes that denature or cause the fragmentation of proteins have been shown to be effective against prions. Moist heat and certain phenolic formulations are proposed to inactivate prions by denaturation. It has been speculated that renaturation of the denatured protein could occur under some conditions, although this has not been observed. Biocides, like sodium hypochlorite, sodium hydroxide, and hydrogen peroxide at high concentrations, appear to cause fragmentation or other structural changes to prions, rendering them non-infectious. It is still possible but considered unlikely that other, as yet unidentified factors are involved in prion infectivity and need to be similarly degraded to ensure complete inactivation.

Almost all inactivation and decontamination procedures to prions reported so far are ineffective. If the procedure is effective to prions, it may cause damage to material and functional compatibility. According to the current sterilization validation and good manufacturing practice (GMP), 10^{-6} SAL (sterility assurance level) and material and functional compatibility must be achieved simultaneously. However, the presented procedures for inactivation of prions and endotoxins are failed to satisfy both SAL of 10^{-6} and material and functional compatibility simultaneously.

Recently gas plasma sterilization was studied (Shintani *et al.*, 2007; Shintani and McDonnell, 2011) and nitrogen gas plasma sterilization can attain both 10^{-6} SAL and material and functional compatibility in success when applied to prion and endotoxin inactivation. The study example is briefly presented below. The pulsed type nitrogen gas plasma generator is very effective to inactivate prion (Fig. 6.2; Shintani *et al.*, unpublished data) by bovine

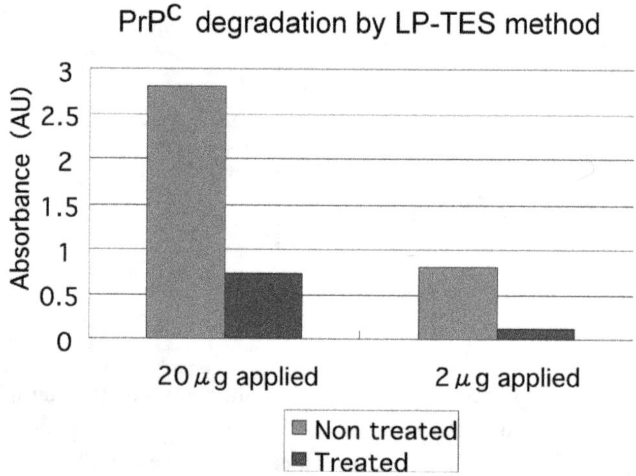

Figure 6.2 20μg or 2μg of PrP^c was applied. After nitrogen gas plasma exposure, the residue PrP^c was sampled with EIA buffer and assayed with an ELISA kit from Funakoshi, Tokyo, Japan. Nitrogen gas plasma exposure condition: (low pressure method) (reactor); anode–cathode gap: 40mm; vacuum: 43,000Pa; nitrogen gas flow speed: 6l/min; time: 30 minutes; temperature: 60°C (temperature was measured inside of plasma); (electric source) input power: 84W; rep. freq. 2.5kHz; peak voltage: 19.0kV. The smaller the PrP^c, the more degradation is observed. This indicates clump phenomenon of PrP^c onto carrier materials can be formed. LP-TES, low-pressure triple effects sterilization; ELISA, enzyme-linked immunosorbent assay; EIA, enzyme immunoassay. A colour version of this figure can be found in the plate section at the back of the book.

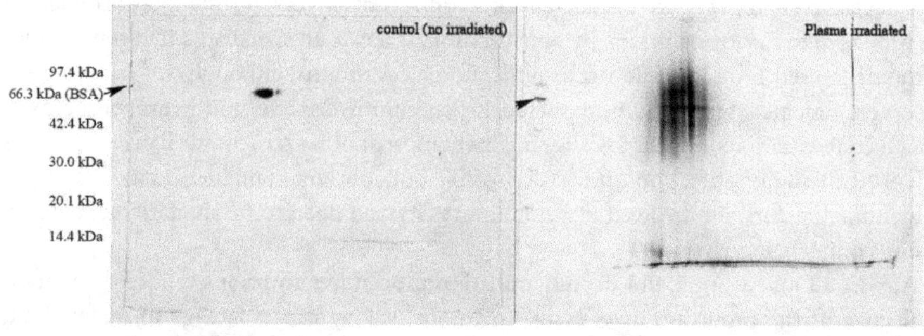

Figure 6.3 Deformation of bovine serum albumin (BSA) by nitrogen gas plasma, which was confirmed by secondary electrophoresis. Each arrow shows the original size of BSA. Compared with the one band of control BSA (left), exposed BSA by nitrogen gas plasma (right) is degraded and fragmentary. Some fragments may occur polymerization. Anyhow, deformation of BSA is occurred by nitrogen gas plasma. Material: bovine serum albumin (BSA) 500ng. Method: secondary SDS polyacrylamide gel electrophoresis. Low pressure nitrogen gas plasma reactor: anode–cathode gap: 40mm; vacuum: 43000Pa; nitrogen gas flow speed: 6l/min; time: 30 minutes; temperature: 60°C (temperature was measured inside of plasma); (electric source) input power: 84W; rep. freq. 2.5kHz; peak voltage:19.0kV. SDS, sodium dodecyl sulfate.

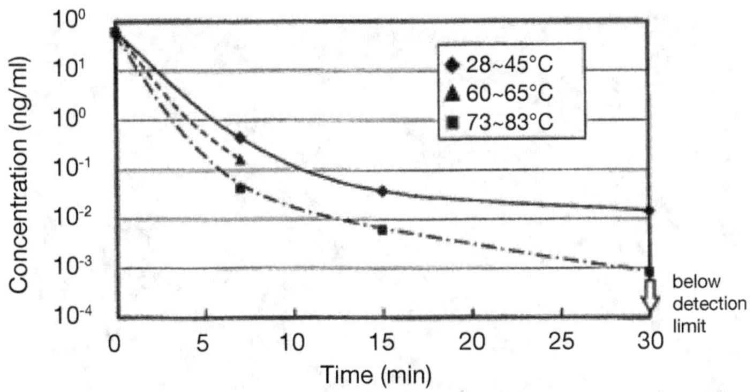

Figure 6.4 Temperature dependence of endotoxin inactivation by low pressure nitrogen gas plasma exposure. Conditions of low pressure nitrogen gas plasma exposure: input power, 84W; frequency, 2.5kHz; temperature, 28–73°C; vacuum, 45kPa; gap, 40mm; flow rate, 15l/min; voltage, 19kV.

serum albumin (BSA; Fig. 6.3; Shintani *et al.*, unpublished data), and endotoxin up to 6 log reduction for 30 minutes exposure (Fig. 6.4; Shintani *et al.*, 2007) with achievement of SAL of 10^{-6} and material and functional compatibility. This is due to the shallow penetration depth of gas plasma sterilization factors such as NO radicals, OH radicals, nitrogen metastables and so on at the 20–40 nm level (Fig 6.5; McDonnell, 2007b; Shintani *et al.*, 2007). This penetration depth is effective to inactivate bioburden including bacterial spores. The shallow penetration depth of nitrogen gas plasma exposure compared with other existing sterilization procedures are the most reason why both SAL of 10^{-6} and material and functional compatibility simultaneously can easily attain. OH radicals bind to amino acids, especially amino acids with benzene ring (phenylalanine, tyrosine, tryptophan), resulting in denaturation of prion protein due to failure of maintenance of high-order structure because of incorrect hydrogen bonding between amino acids.

USFDA requires at least three log reduction of endotoxin (FDA Guidance for Industry, 2004) and dry heating at 250°C for 30 minutes is recommended, but almost all polymer materials cannot withstand by this procedure. Three log reduction can be achieved by using nitrogen gas plasma exposure for 10 minutes at 28–45°C without any deterioration of materials (Fig. 6.4; Shintani *et al.*, 2007). More than 5 log reduction of endotoxins was completed by nitrogen gas plasma exposure at 73–83°C for 30 minutes together with material and functional compatibility and 10^{-6} SAL (Fig. 6.4; Shintani *et al.*, 2007). Inactivation of endotoxin can be conducted by the ester bond cleavage of fatty acid of Lipid A (Williams, 2007), active site of endotoxin, by OH and/or NO radicals produced from nitrogen gas plasma exposure. As a result, m/z of 1450 and/or 1435, which cleaved one ester bonding of Lipid A of *E. coli* respectively, was identified (unpublished data by Shintani). The pyrogenic activity of m/z of 1450 and 1435 was confirmed to be lost.

From these results, we can confirm nitrogen gas plasma exposure is effective to attain prion and endotoxin inactivation. In addition both SAL of 10^{-6} and material/functional compatibility can be successfully achieved.

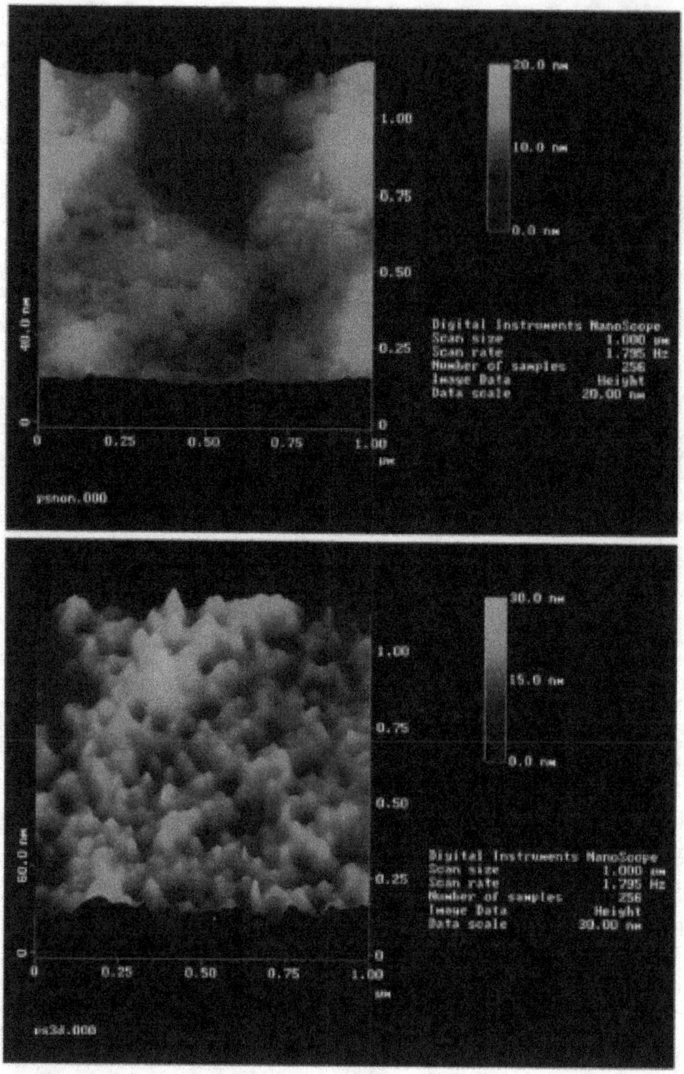

Figure 6.5 Surface analysis of atomic force microscopy before and after nitrogen gas plasma exposure to polystyrene. Lower figure: before gas exposure (control); upper figure: after gas plasma treatment for 7 minutes by low pressure.

References

Fairand, B.P. (2002). Radiation Sterilization for Healthcare Products (CRC Press, New York), pp. 39–41.

Fichet, G., Comoy, E., Guvol, C., Antloga, K., Dehen, C., Charbonnier, A., McDonnell, G., Brown, P., Lasmezas, C., and Deslys, J.-P. (2004). Novel methods for disinfection of prion-contaminated medical devices. Lancet *364*, 521–526.

Halliwell, G., and Gutteridge, J.M.C. (2007). Free Radicals in Biology and Medicine (Oxford University Press, New York), pp. 222–229.

McDonnell, G.E. (2007a). Mechanisms of Prion Resistance in Antisepsis, Disinfection, and Sterilization (ASM Press, Washington DC), pp. 318–320.

McDonnell, G.E. (2007b). Plasma in Antisepsis, Disinfection, and Sterilization (ASM Press, Washington DC), pp. 184–186.

Shintani, H., and McDonnell, G. (2011). Inactivation of microorganisms (spore types and vegetative cells) and the mechanism by gas plasma. In Sterilization and Disinfection by Plasma, Sakudo, A., and Shintani, H., eds. (Nova Science Publishers, Inc., New York), pp. 33–48.

Shintani, H., Shimizu, N., Imanishi, Y., Sekiya, T., Tamazawa, K., Taniguchi, A., and Kido, N. (2007). Inactivation of microorganisms and endotoxin by low temperature nitrogen gas plasma exposure. Biocontrol Science *12*, 131–143.

Taylor, D.M. (2004). Transmissible degenerative encephalopathies: inactivation of the unconventional causal agents. In Principles and Practice of Disinfection Preservation and Sterilization, Fraise, A.P., Lambert, P.A., and Maillard, J.-Y., eds. (Blackwell publishing, Oxford, UK), pp. 324–344.

Williams, L.K. (2007). Microbial biodiversity and lipopolysaccharide heterogeneity: from static to dynamic models. In Endotoxins, Pyrogens, LAL Testing and Depyrogenetion (Informa Healthcare, New York), pp. 111–131.

Clinical Aspects of Human Prion Diseases

7

Richard Knight

Abstract

Human prion disease is divided into three broad classes: idiopathic, genetic and acquired. There are significant differences in clinical presentation both between and within these three groups, but all are progressive, fatal brain diseases with dementia, cerebellar ataxia and involuntary movements being particularly prominent features. Absolutely definite diagnosis requires neuropathological analysis of brain tissue but there are established clinical diagnostic criteria and a variety of supportive investigations including abnormalities in the EEG, cerebral MRI and CSF protein analysis. For variant CJD, tonsil biopsy is an additional test and, in genetic prion disease, blood testing for pathogenic *PRNP* mutations is possible.

Introduction

Human prion disease is divided into three broad classes, principally according to causation: Idiopathic, Genetic and Acquired (Table 7.1). This rather heterogeneous grouping is unified by a common neuropathological appearance, molecular underpinning and potential

Table 7.1 The present classification of human prion diseases

Causation category	Specific forms	Comments
Idiopathic	Sporadic CJD	Worldwide distribution
	Variably protease sensitive prionopathy	Precise status of VPSPr uncertain
Genetic	Genetic CJD, Gerstmann–Sträussler–Scheinker syndrome, FFI	Many recognized *PRNP* pathogenic mutations Three forms defined according to clinicopathological phenotype Autosomal dominant inheritance pattern
Acquired	Kuru	Confined to Papua New Guinea
	Iatrogenic CJD	Classically related to cadaveric derived pituitary hormones and dura mater grafts or corneal transplantation/depth EEG electrodes/neurosurgery
	Variant CJD	Primary infection from BSE contamination of food. Secondary cases from blood transfusion and use of blood products. Most cases in the UK and France

transmissibility. The clinical features vary partly with causation and partly with other factors. However, they are all characterized by a progressive encephalopathy, usually with prominent dementia, that is universally eventually fatal (with no current proven effective treatment). Absolutely definite diagnosis depends on neuropathological examination of brain tissue; if this is done, it is usually at autopsy but there are occasional indications for brain biopsy. Standardized clinical diagnostic criteria have been developed and are employed in international collaborative surveillance and research, also being adopted by the WHO (www.cjd.ed.ac.uk/criteria.htm). There are five clinical tests that may be helpful, depending on the situation: Brain MRI, EEG, CSF protein analysis, tonsil biopsy (in variant CJD) and *PRNP* genetic testing. There are no abnormal non-neurological signs in human prion disease and routine haematological and biochemical blood tests are usually normal. Typically, the CSF is acellular with a normal glucose level but, rarely, excess white cells are seen and the CSF total protein level may be raised. Cell counts between 5 and 20 could be seen in human prion disease but should prompt serious consideration of other diagnoses. In a review of CSF tests in prion disease, Green *et al.* found that a CSF white cell count of >20 and/or a CSF total protein level of >1 g/l strongly suggest an alternative diagnosis (Green *et al.*, 2007). In the same review, they described a few cases with findings of CSF oligoclonal IgG, but this is unusual. Brain imaging, especially CT, may be normal and the contrast between a seriously neurologically disabled person and the normality of brain imaging may prompt consideration of the diagnosis. In the discussion below, clinical tests are considered from the point of view of abnormalities that may positively suggest or support the diagnosis of prion disease, however they have important roles in the exclusion or confirmation of alternative conditions.

Idiopathic human prion disease

The commonest human prion disease, sporadic CJD (sCJD), remains of unknown cause. It has a worldwide distribution and occurs in all *PRNP*-129 genotypes, although more commonly in MM and less commonly in MV individuals than expected by chance. The majority of cases follow a reasonably uniform clinical course, but there is a significant degree of clinical (and pathological) heterogeneity of uncertain significance.

A new clinicopathological phenotype of human prion disease was described in 2008, in 11 patients identified in the USA, and originally named 'protease sensitive prionopathy' (PSPr) (Gambetti *et al.*, 2008). It was associated with the accumulation of abnormal prion protein, although one that was significantly more sensitive to protease digestion than typically found in sCJD. No known risk factors for iatrogenic CJD and no pathogenic *PRNP* mutations were found in these cases but all had the *PRNP*-129 VV genotype. A later publication from the same centre described further cases but with a variety of *PRNP*-129 genotypes patients and varying degrees of protease sensitivity; the name 'variably protease sensitive proteinopathy' (VPSPr) was therefore proposed (Zou *et al.*, 2010). Similar cases have been reported from the UK, the Netherlands and Spain (Head *et al.*, 2009, 2010; Jansen *et al.*, 2010; Rodriguez-Martinez *et al.*, 2010). There is evidence that VPSPr is transmissible: a recent report described its successful transmission to bank voles (Nonno *et al.*, 2012). The nosological status of these cases remains uncertain but it is reasonable, for the moment, to regard them as forms of sporadic human prion disease that have clinicopathological differences from sCJD. Of course, sCJD in general shows some clinicopathological heterogeneity

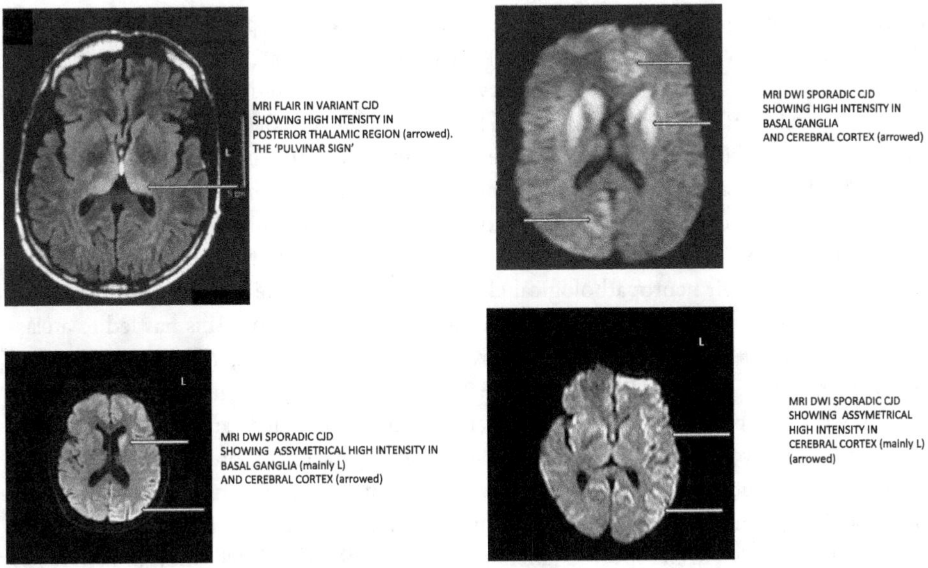

Figure 7.1 The cerebral MR in sCJD and vCJD. Images courtesy of Dr David Summers.

but the molecular characteristics of the described VPSPr cases suggest it is reasonable to consider them separately at this time.

Sporadic CJD

Clinical features

This is a disease predominantly of the middle aged and elderly, although relatively young cases are very occasionally seen. Typically, it is a rapidly progressive illness: in many studies, the median duration is around 4 months. Relatively long survival is seen in a small percentage of cases and, of course, supportive medical and nursing care can prolong life. A collaborative European study identified four independent factors that affect survival: sex, age at onset, *PRNP* codon 129 genotype and PrP^{Sc} protein type (Pocchiari *et al.*, 2004).

Around three-quarters of cases of sCJD follow a fairly uniform clinical course, with rapid progression and multifocal or generalized brain features; typically, dementia, cerebellar ataxia, visual disturbance and myoclonus are dominant features. However, there are other disease patterns and sCJD shows significant clinicopathological heterogeneity (with variable clinical features being reflected in variations in the neuropathological characteristics). Two main variations are presentation with isolated neurological deficits and cases with atypical overall clinical phenotype.

Those presenting with isolated deficits can cause particular diagnostic difficulty. Two well-defined instances are presentation with a pure cerebellar ataxia (sometimes called the 'Brownell–Oppenheimer' variant of sCJD) and a progressive pure visual loss (sometimes called the 'Heidenhain' variant of sCJD). Cerebellar ataxia can be present as an isolated deficit for a few weeks to even a few months before other features emerge and there are many possible causes of progressive ataxia, many being commoner than sCJD (Cooper *et al.*, 2006). Cerebellar onset cases usually progress more slowly than typical sCJD and have overall a longer survival. Pure visual loss may lead to an initial ophthalmological referral, though its characteristics are suggestive of visual cortex disturbance. It typically progresses to cortical blindness and often does so very rapidly with a subsequent clinical course like that of typical sCJD (Cooper *et al.*, 2005). Variations in the overall clinical picture are uncommon but a relatively gradually progressive dementia picture can be seen and an illness dominated by sleep and autonomic disturbances is recognized (sometimes termed 'sporadic fatal insomnia' because of its resemblance to Fatal Familial Insomnia).

The significance of these phenotypic variations remains uncertain but the clinical profiles (along with their neuropathological characteristics) correlate broadly with molecular factors: the *PRNP* codon 129 genotype and the PrP^{Sc} protein type. This has led to a classification of sCJD using the genotype/protein type combination as a base -hence 'MMI sCJD or 'MVII' sCJD etc. (Parchi *et al.*, 1999). The 'sporadic fatal insomnia' phenotype mentioned above being classed as MMII-thalamic on this scheme (Parchi *et al.*, 1999).This system is generally useful in diagnosis and overall case characterization but it is based on an association (a clinicopathological-molecular one) and its precise meaning or significance is not clear. Interestingly, a correlation between the molecular factors and the pattern of brain MR findings has also been reported (Meissner *et al.*, 2009). The distribution of *PRNP*-129 genotypes in normal populations has been studied, showing an east–west variation (Nurmi *et al.*, 2003). It has become increasingly clear that co-occurrence of both protein types (I and II) is found in a proportion of sCJD brains, leading to attempts to refine the classification system but there are obvious remaining uncertainties (Puoti *et al.*, 1999, Head *et al.*, 2004, Cali *et al.*, 2009, Parchi *et al.*, 2009).

Diagnosis

The absolutely definite diagnosis of sCJD requires neuropathological examination of brain tissue, usually obtained at autopsy; brain biopsy may be indicated occasionally in specific circumstances, for example, the need to exclude an alternative, treatable disease. There is no validated simple clinical diagnostic test, but there are useful supportive investigations. The diagnostic process involves three main steps: suspecting sCJD on the basis of a suggestive clinical profile, excluding other possible diagnoses and finding certain abnormalities in the EEG, cerebral MRI and CSF protein testing.

The EEG becomes progressively abnormal in sCJD and, in a significant number of cases, characteristic generalized periodic discharges eventually develop. The proportion of cases that develop this EEG pattern depends somewhat on the EEG testing policy adopted; repeated recordings increase the number of positive results since the characteristic pattern may develop with clinical progression, if it is not initially present. Repeat EEG recordings at around weekly intervals may be helpful. In one study the sensitivity of the EEG as a test for sCJD was 44%,with a specificity of 92% (Zerr *et al.*, 2009). In a significant majority of cases, the cerebral MRI shows characteristic signal change in the putamen and caudate and/or areas of the cerebral cortex (Collie *et al.*, 2001; Fig. 7.1). One review reported a sensitivity and specificity both of 83% for the MRI as a sCJD test (Zerr *et al.*, 2009). As noted above, the MR findings in sCJD may vary with the *PRNP*-129 genotype (Meissner *et al.*, 2009). The changes in the basal ganglia have a higher specificity for sCJD than the cortical changes and the presently accepted international diagnostic criteria embody only the former (www.cjd.ed.ac.uk/criteria.htm). Analysis of the CSF may show elevated levels of several proteins but 14-3-3 and S100b have proved most useful in practice. A positive CSF 14-3-3 test was reported to have a sensitivity and specificity for sCJD of 86% and 68% respectively (Zerr *et al.*, 2009). In a review of a 10-year experience with CSF protein testing in UK CJD surveillance, the sensitivity and specificity of a positive CSF 14-3-3 test for sCJD were 86% and 74% respectively. A combination of a positive 14-3-3 test and an elevated S100b level had a sensitivity and specificity of 95% and 94% respectively (Chohan *et al.*, 2012).

The findings described here in all three tests are not absolutely specific for sCJD and they need to be viewed in the individual clinical context. The quoted specificities will reflect the particular defined testing context and are not necessarily valid if tests are done in different contexts or in line with a different testing policy. A recently reported test (real-time quaking-induced conversion or RTQuIC), using CSF, employs a methodology linked to the specific molecular basis of prion disease: the abnormality in the prion protein. If PrP^{Sc} is present in CSF in sCJD, it is present in an amount difficult to detect using current detection methods. The basis of the RTQuIC test is an amplification technique that aims to increase PrP^{Sc} levels to ones that are readily detectable (Atarashi *et al.*, 2011). The test has proved promising in sCJD and is currently being evaluated in the UK surveillance system (Peden *et al.*, 2009; McGuire *et al.*, 2012).

Variably protease sensitive prionopathy (VPSPr)

Clinical features

The overall picture is that of a progressive brain disease in middle aged to elderly individuals but with a generally slower pace of progression and a longer duration than that typical of sCJD. In their description of 26 cases, Zou *et al.* reported a mean age at onset of 67 years (range 48–81 years) and a mean illness duration of 30 months (Zou *et al.*, 2010). They reported a higher incidence of psychiatric and parkinsonian features, and a lower incidence of myoclonus, than found in sCJD. Cognitive decline was usually relatively slow compared with that typically seen in sCJD. A single case report, from the Netherlands, described a 54-year-old with an illness duration of 20 months and a clinical picture involving dementia, spastic paraplegia and a sensorimotor polyneuropathy (Jansen *et al.*, 2010). Two cases have been reported from the UK, a 56-year-old and a 76-year-old, with illness durations of 42 and 12 months respectively (Head *et al.*, 2009, 2010). In one of the UK cases, the clinical presentation was a dementia with frontal lobe features (Head *et al.*, 2009).

Given the relatively small number of described cases and variations in the features reported, there must be some uncertainty as to the precise clinical profile of this disease.

Diagnosis

The recognition of these cases is potentially problematic as they appear to present with a variable clinical picture suggestive of some sort of neurodegenerative disease in the middle-aged and elderly, with features that could suggest other more common diagnoses. In the report from Zou *et al.*, a positive CSF 14-3-3 was found in only 2/12 cases; a periodic EEG pattern in only 1/18 and brain MRI changes similar to those seen in sCJD in only 1/25 cases (Zou *et al.*, 2010). No *PRNP* mutations have been reported in these cases and all *PRNP* codon 129 genotypes have been affected. The diagnosis therefore rests on the neuropathological findings along with the characteristic prion protein characteristics (Gambetti *et al.*, 2008; Zou *et al.*, 2010). However, it is not likely that all cases would come to autopsy and, even if they do, the appropriate neuropathological and molecular studies need to be undertaken.

Genetic prion disease

Clinical features

Genetic prion diseases (gPD) result from *PRNP* pathogenic mutations, which are inherited on an autosomal dominant basis; many are now described but they fall into two broad groups: point mutations and insertional mutations (the latter being found in an octapeptide repeat section) (Kovacs *et al.*, 2002, 2005).

These illnesses have, overall, a younger age of onset, and a longer duration, than for sCJD, but they can present relatively late in life and can have a rapid progression (Kovacs *et al.*, 2005). The clinical features vary significantly, according partly (but not entirely) to the particular underlying mutation and the *PRNP* 129 genotype. (Kovacs, 2002). The genetic prion diseases have been divided into genetic CJD (gCJD), Gerstmann–Sträussler–Scheinker syndrome (GSS) and fatal familial insomnia (FFI) according to cliniopathological phenotype. The original description of GSS was essentially that of a familial progressive cerebellar ataxia (with other features) and typically associated with the *PRNP* P102L mutation but a number of different clinical pictures and underlying mutations have been identified. The term 'GSS' has now become associated with a characteristic neuropathological picture involving multicentric prion plaques in the brain (Bugiani *et al.*, 2000). FFI, associated with the *PRNP* D178N mutation, has a particular clinical profile involving sleep disturbance and autonomic features with particular neuropathological findings (Medori *et al.*, 1992).

The illness phenotype of gPD may be indistinguishable from that of sCJD (and, sometimes, other prion diseases). It is, arguably, time to leave aside the eponymous or phenotypic terminology and refer to the group simply as gPD and then specifying the associated mutation. Although these diseases are inherited on an autosomal dominant basis, a family history is absent in around 47% of cases overall, varying with particular mutation (Kovacs *et al.*, 2005).

Diagnosis

Genetic testing for known *PRNP* pathogenic mutations can be undertaken on blood. An absolutely definitive diagnosis is made by neuropathological findings and the demonstration of a known pathogenic mutation. The initial diagnosis rests principally on a clinical suspicion, based on the clinical picture, with, in many cases, a family history of either undiagnosed neurodegenerative disease or proven prion disease. The EEG, CSF and cerebral MRI may show abnormalities similar to those seen in sCJD, especially in E200K cases, the commonest *PRNP* mutation disease, but generally less often (Kovacs *et al.*, 2002, 2005). The fact that a family history may be absent in up to 47% of cases overall means that genetic testing is necessary to absolutely exclude a genetic prion disease. It is reasonable to consider *PRNP* mutation testing in a variety of progressive brain disease settings if no other diagnosis is obvious.

Acquired human prion disease

The clinical features of acquired human prion disease vary according to cause. However, the illness may be indistinguishable from genetic or sporadic disease and, therefore, the final diagnostic classification may depend on historical information identifying a recognized risk exposure. Known causes of iatrogenic CJD should be enquired about in the clinical evaluation of a case of suspected prion disease. Having said that, certain potentially relevant events, for example ophthalmic surgery or blood transfusion, are relatively frequent procedures; in the absence of any additional specific fact – such as the relevant donor having had CJD – the preceding event might be simply coincidental.

Kuru, with its particular geographical distribution and its current extreme rarity, will not be described here but generally presented with a progressive cerebellar ataxia; cognitive impairment occurring relatively late in the illness.

Variant CJD

Clinical features

Most cases of variant CJD (vCJD) have occurred in the UK, with a number in France and a few cases in other countries (Table 7.2). This geographical distribution is important in considering newly suspected vCJD cases. Atypical sCJD can clinically resemble vCJD, and in a country that has never previously reported vCJD, the former is a more likely diagnosis than the latter. Naturally, this does not mean that vCJD should not be considered and some cases attributed to some countries are thought to have contracted the disease through time spent in the UK.

The age at onset of disease is significantly lower than that of sCJD, with a longer illness duration (median durations of, respectively, around 14 and 4 months in the UK).

The typical presentation is one with a psychiatric profile, with definitively neurological features typically occurring later and often clear-cut abnormal neurological signs being found only after a few months. Some relatively early features that are later identifiable as neurological problems are non-specific (such as pain, dizziness, poor concentration or visual blurring) and are easily initially attributable to psychiatric disturbance or to side effects of

Table 7.2 Variant CJD cases (as of 13.09.12) (cases classified as definite or probable according to WHO accepted diagnostic criteria)

Country	Number of cases	Comments
UK	176	173 primary and three secondary (blood transfusion)
France	27	
Republic of Ireland[a]	4	One case considered to have contracted vCJD in the UK
Italy	2	
USA[a]	3	None considered not to have contracted vCJD in the USA
Canada[a]	2	One case considered to have contracted vCJD in the UK
Saudi Arabia	1	
Japan[a]	1	Considered by Japan to be likely to have contracted vCJD in the UK
Netherlands	3	
Portugal	2	
Spain	5	
Taiwan[a]	1	Considered likely to have contracted vCJD in the UK

Cases attributed to a country according to normal residence at time of diagnosis. This is not necessarily the country where the disease was contracted. Data available at: www.cjd.ed.ac.uk.
[a]Countries with reported cases that had residence in either the UK or Saudi Arabia that is considered relevant to causation.

drug treatments (such as antidepressants) given. An analysis of the first hundred cases in the UK concluded that early differentiation of a primary psychiatric illness (such as depression) and the presentation of vCJD is very problematic; early diagnosis may be impossible and, of course, conditions such as depressive illness are very much commoner than vCJD (Spencer *et al.*, 2002). The passage of time may, therefore, be an integral part of the diagnostic process-as it is in a number of neurological conditions. Definite cognitive impairment develops in all cases and cerebellar ataxia in nearly all (Heath *et al.*, 2010). Persistent painful or unpleasant sensory symptoms occur in around two thirds of cases (Macleod *et al.*, 2002, Heath *et al.*, 2010). Chorea, dystonia or myoclonus is seen in the majority of cases (Heath *et al.*, 2010). The illness progresses to one of increasing neurological disability and, in longer surviving cases, an eventual state of immobility and mutism.

Secondary vCJD (resulting from blood transfusion or blood products) is clinically indistinguishable from primary (diet-related) vCJD.

Diagnosis

The differential diagnosis of a progressive neuropsychiatric illness in a young person is potentially very wide; the initial differential diagnoses made by neurologists in the first 150 UK cases have been reviewed (Heath *et al.*, 2011). The CSF 14-3-3 test is significantly less sensitive for vCJD than it is for sCJD. Testing the CSF for phosphorylated tau is more helpful (Goodall *et al.*, 2006). The EEG may be normal in the early stages and typically shows only non-specific slow wave activity with disease progression. The periodic discharges characteristic of sCJD are usually not seen in vCJD although there are two case reports where they have appeared very late in the illness (Binelli *et al.*, 2006; Yamada *et al.*, 2006). The cerebral MRI is a very helpful investigation and the characteristic finding in vCJD is high signal in the pulvinar region of the thalamus which is found in over 90% of cases, depending in part on the MR sequences employed, FLAIR and DWI having the greatest sensitivity (Collie *et al.*, 2001; Zeidler *et al.*, 2001; Collie *et al.*, 2003; Fig. 7.1). In vCJD, in contradistinction to other forms of human prion disease, abnormal prion protein (PrP^{Sc}) may be found outside the central nervous system, in lymphoreticular tissues. As a result, a tonsil biopsy, with examination for protease resistant prion protein, is a potentially very useful test for vCJD (Hill *et al.*, 1997). It is, of course, somewhat invasive, with potential morbidity, and is arguably best reserved for those cases where significant doubt about the diagnosis exists (for example, in those cases with rather atypical clinical profiles or where the cerebral MRI has not shown the pulvinar abnormality). In considering vCJD as a diagnosis, it is important to bear in mind its great rarity (even in the UK, over the 3-year period 2009–2011, only nine diagnoses of vCJD were made) and its geographical distribution (Table 7.2; www.cjd.ac.uk). An additional important consideration is that all clinical cases to date classed as definite or probable according to the established diagnostic criteria have been in *PRNP* 129 MM individuals, although it is known that infection can occur in MV or VV individuals and there is a report of a likely MV clinical case (Kaski *et al.*, 2009).

Iatrogenic CJD

Clinical features

Secondary vCJD is, in essence, an iatrogenic illness, but the term 'iatrogenic CJD' is still generally used to refer to presumed or proven transmissions of sporadic (or possibly genetic)

CJD. The worldwide occurrence of iCJD was reviewed in 2006 (Brown *et al.*, 2006). In general, cases of iatrogenic CJD (iCJD) that result from instances other than human growth hormone, have a clinical picture like that of sCJD and it is the history of prior exposure (for example via cadaveric-derived human dura mater graft) that indicates the diagnosis.

The characteristics of cases resulting from the surgical use of cadaveric-derived dura mater have been reviewed in Japan (where the majority of cases have occurred) and in the UK (Heath *et al.*, 2006; Noguchi-Shinohara *et al.*, 2007).

Cases due to cadaveric-derived human growth hormone tend to present with a progressive cerebellum ataxia, with relatively mild or late cognitive impairment (thereby resembling Kuru) (Brown *et al.*, 1992).

Diagnosis

The diagnosis of iatrogenic CJD depends largely on the consideration of prion disease with a history of a known relevant exposure to a risk factor. In the case of exposure to cadaveric-derived human growth hormone or gonadotropin, the exposure should be known and, indeed, individuals at such risk have been informed by the relevant health bodies. Past use of a human dura mater graft may not be so obvious and a history of previous potentially relevant surgery will need exploration. Other causes are very rare. The significance of previous ophthalmological surgery, not uncommon in the general population, may be difficult to assess. The length of time between the putative exposure and the illness onset is a relevant factor, bearing in mind the relatively long incubation periods in acquired prion disease (Brown *et al.*, 2006). The relatively distinctive picture seen in human growth hormone cases is noted above although, of course, there are various possible causes of a progressive cerebellar ataxic syndrome.

References

Binelli, S., Agazzi, P., Giaccone, G., Will, R.G., Bugiani, O., Franceschetti, S., and Tagliavinin, F. (2006). Periodic electroencephalogram complexes in a patient with variant Creutzfeldt–Jakob disease. Ann. Neurol. *59*, 423–427.

Brown, P., Preece, M.A., and Will, R.G. (1992). 'Friendly fire' in medicine: hormones, homografts, and Creutzfeldt–Jakob disease. Lancet *340*, 24–27.

Brown, P., Brandel, J.-P., Preece, M., and Sato, T. (2006). Iatrogenic Creutzfeldt–Jakob Disease. The waning of an era. Neurology *67*, 389–393.

Bugiani, O., Giaccone, G., Piccardo, P., Morbin, M., Tagliavini, F., and Ghetti, B. (2000). Neuropathology of Gerstmann-Sträussler-Scheinker Disease. Microsc. Res. Tech. *50*, 10–15.

Cali, I., Castellani, R., Alshekhlee, A., Cohen, Y., Blevins, J., Yuan, J., Langeveld, J.P., Parchi, P., Safar, J.G., Zou, W.Q., *et al.* (2009). Co-existence of scrapie prion protein types 1 and 2 in sporadic Creutzfeldt–Jakob disease: its effect on the phenotype and prion-type characteristics. Brain *132(Pt 10)*, 2643–2658.

Chohan, G., Pennington, C., Mackenzie, J.M., Andrews, M., Everington, D., and Will, R.G. (2010). The role of cerebrospinal fluid 14–3–3 and other proteins in the diagnosis of sporadic Creutzfeldt–Jakob disease in the UK: a 10-year review. JNNP *81*, 1243–1248.

Collie, D.A., Sellar, R.J., Zeidler, M., Colchester, A.F.C., Knight, R., and Will, R.G. (2001). MRI of Creutzfeldt–Jakob disease: imaging features and recommended MRI protocol. Clin. Radiol. *56*, 726–739.

Collie, D.A., Summers, D.M., Sellar, R.J., Ironside, J.W., Cooper, S., Ziedler, M., Knight, R., and Will, R.G. (2003). Diagnosing variant Creutzfeldt–Jakob disease with the pulvinar sign: MR imaging findings in 86 neuropathologically confirmed cases. Am. J. Neuroradiol. *24*, 1560–1569.

Cooper, S.A., Murray, K.L., Heath, C.A., Will, R.G., and Knight, R.S.G. (2005). Isolated visual symptoms at onset in sporadic Creutzfeldt–Jakob disease: the clinical phenotype of the 'Heidenhain variant'. Br. J. Ophthalmol. *89*, 1341–1342.

Cooper, S.A., Murray, K.L., Heath, C.A., Will, R.G., and Knight, R.S.G. (2006). Sporadic Creutzfeldt–Jakob disease with cerebellar ataxia at onset in the United Kingdom. JNNP *77*, 1273–1275.

Gambetti, P., Dong, Z., Yuan, J., Xiao, X., Zheng, M., Alshekhlee, A., Castellani, R., Cohen, M., Barria, M.A., Gonzalez-Romero, D., *et al.* (2008). A novel human disease with abnormal prion protein sensitive to protease. Ann. Neurol. *63*, 697–708.

Goodall, C.A., Head, M.W., Everington, D., Ironside, J.W., Knight, R.S., and Green, A.J. (2006). Raised CSF phospho-tau concentrations in variant CJD: diagnostic and pathological implications. J. Neurol. Neurosurg. Psychiatry *77*, 89–91.

Green, A.J.E., Thompson, E.J., Stewart, G.E., Ziedler, M., McKenzie, J.M., MacLeod, M.A., Ironside, J.W., Will, R.G., and Knight, R.S. (2001). Use of 14-3--3 and other brain-specific proteins in CSF in the diagnosis of variant Creutzfeldt–Jakob disease. JNNP *70*, 744–748.

Green, A., Sanchez-Juan, P., Ladogana, A., Cuadrado-Corrales, N., Sáanchez-Valle, R., Mitrováa, E., Stoeck, K., Sklaviadis, T., Kulczycki, J., Heinemann, U., *et al.* (2007). CSF analysis in patients with sporadic CJD and other transmissible spongifrm encephalopathies. Eur. J. Neurol. *14*, 121–124.

Head, M.W., Bunn, T.J.R., Bishop, M.T., McLoughlin, V., Lowrie, S., McKimmie, C.S., Willliams, M.C., McCardle, L., Mackenzie, J., Knight, R., *et al.* (2004). Prion protein heterogeneity in sporadic but not variant Creutzfeldt–Jakob disease: UK cases 1991–2002. Ann. Neurol. *55*, 851–859.

Head, M.W., Knight, R., Zeidler, M., Yull, H., Barlow, A., and Ironside, J.W. (2009). A case of protease-sensitive prionopathy in a patient in the United Kingdom. Neuropathol. Appl. Neurobiol. *35*, 628–632.

Head, M.W., Lowrie, S., Chohan, G., Knight, R., Scoones, D.J., and Ironside, J.W. (2010). Variably protease-sensitive prionopathy in a *PRNP* codon 129 heterozygous UK patient with co-existing tau, synuclein and A beta pathology. Acta Neuropathol. *120*, 821–823.

Heath, C.A., Barker, R.A., Esmonde, T.F.G., Harvey, P., Roberts, R., Trend, P., Head, M.W., Smith, C., Bell, J.E., Ironside, J.W., *et al.* (2006). Dura mater-associated Creutzfeldt–Jakob disease – experience from surveillance in the UK. J. Neurol. Neurosurg. Psychiatry *77*, 880–882.

Heath, C.A., Cooper, S.A., Murray, K., Lowman, A., Henry, C., MaCleod, M.A., Stewart, G.E., Zeidler, M., MacKenzie, J.M., Ironside, J.W., *et al.* (2010). Validation of diagnostic criteria for variant CJD. Ann. Neurol. *67*,761–770.

Heath, C.A., Cooper, S.A., Murray, K., Lowman, A., Henry, C., MacLeod, M.A., Stewart, G., Zeidler, M., McKenzie, J.M., Knight, R.S., *et al.* (2011). Diagnosing variant Creutzfeldt–Jakob disease: a retrospective analysis of the first 150 cases in the UK. JNNP *82*, 646–651.

Hill, A.F., Zeidler, M., Ironside, J., and Collinge, J. (1997). Diagnosis of new variant Creutzfeldt–Jakob disease by tonsil biopsy. Lancet *349*, 99–100.

Jansen, C., Head, M.W., van Gool, W.A., Bass, F., Yull, H., Ironside, J.W., and Rozemuller, A.J.M. (2010). The first case of protease sensitive prionopathy (PSPr) in the Netherlands: a patient with an unusual GSS-like clinical phenotype. JNNP *81*, 1052–1055.

Kaski, D., Mead, S., Hyare, H., Cooper, S., Jampana, R., Overell, J., Knight, R., Collinge, J., and Rudge (2009). Variant CJD in an individual heterozygous for *PRNP* codon 129. Lancet *374*, 2128.

Kovacs, G.G., Trabattoni, G., Hainfellner, J.A., Ironside, J.W., Knight, R.S.G., and Budka, H. (2002). Mutations of the prion protein gene: phenotypic spectrum. J. Neurol. *249*, 1567–1582.

Kovacs, G.G., Puopolo, M., Ladogana, A., Pocchiari, M., Budka, H., van Duijin, C., Collins, S.J., Boyd, A., Giulivi, A., Coulthart, M., *et al.* (2005). Genetic prion disease: the EUROCJD experience. Hum. Genet. *118*, 166–174.

Macleod, M.A., Stewart, G., Zeidler, M., Will, R.G., and Knight, R. (2002). Sensory features of variant Creutzfeldt–Jakob disease. J. Neurol. *249*, 706–711.

McGuire, L.I., Peden, A.H., Orrú, C.D., Wilham, J.M., Appleford, N.E., Mallinson, G., Andrews, M., Head, M.W., Caughey, B., Will, R.G., *et al.* (2012). Real time quaking-induced conversion analysis of cerebrospinal fluid in sporadic Creutzfeldt–Jakob disease. Ann. Neurol. *72*, 278–285.

Medori, R., Tritschler, H.J., LeBlanc, A., Villare, F., Manetto, V., Ying Chen, H., Xuf, R., Leal, S., Montagna, P., Cortelli, P., *et al.* (1992). Fatal familial insomnia, a prion disease with a mutation at codon 178 of the prion protein gene. New Engl. J. Med. *326*, 444–449.

Meissner, B., Kallenberg, K., Sanchez-Juan, P., Collie, D., Summers, D.M., Almonti, S., Collins, S.J., Smith, P., Cras, P., Jansen, G.H., *et al.* (2009). MRI lesion profiles in sporadic Creutzfeldt–Jakob disease. Neurology *72*, 1994–2001.

Noguchi-Shinohara, M., Hamaguchi, T., Kitamoto, T., Sato, T., Nakamur, Y., Mizusawa, H., and Yamada, M. (2007). Clinical features and diagnosis of dura mater associated Creutzfeldt–Jakob disease. Neurology *69*, 360–367.

Nurmi, M.H., Bishop, M., Strain, L., *et al.* (2003). The normal population distribution of *PRNP* codon 129 polymorphism. Acta Neurol. Scand. *108*, 374–378.

Parchi, P., Giese, A., Capellari, S., Brown, P., Schulz-Schaeffer, W., Windl, O., Zerr, I., Budka, H., Kopp, N., Piccardo, P., *et al.* (1999). Classification of sporadic Creutzfeldt–Jakob disease based on molecular and phenotypic analysis of 300 subjects. Ann. Neurol. *46*, 224–233.

Parchi, P., Strammiello, R., Notari, S., Giese, A., Langeveld, J.P.M., Ladogana, A., Zerr, I., Roncaroli, F., Cras, P., Ghetti, B., *et al.* (2009). Incidence and spectrum of sporadic Creutzfeldt–Jakob disease variants with mixed phenotype and co-occurrence of PrPSc types: an updated classification. Acta Neuropathol. *118*, 659–671.

Pocchiari, M., Puopolo, M., Croes, E.A., Budka, H., Gelpi, E., Collins, S., Lewis, V., Sutcliffe, T., Giulivi, A., Delasnerie-Laupretre, N., *et al.* (2004). Predictors of survival in sporadic Creutzfeldt–Jakob disease and other human transmissible spongiform encephalopathies. Brain *127*, 2348–2359.

Puoti, G., Giaccone, G., Rossi, G., Canciani, B., Bugiani, O., and Tagliavini, F. (1999). Sporadic Creutzfeldt–Jakob disease: co-occurrence of different types of PrPsc in the same brain. Neurology *53*, 2173–2176.

Spencer, M.D., Knight, R.S.G., and Will, R.G. (2002). First hundred cases of variant Creutzfeldt–Jakob disease: retrospective case note review of early psychiatric and neurological features. BMJ *324*, 1479–1482.

Yamada, M. (2006). The first Japanese case of variant Creutzfeldt–Jakob disease showing periodic electroencephalogram. Lancet *367*, 874.

Zeidler, M., Collie, D.A., Macleod, M.A., Sellar, R.J., and Knight, R. (2001). FLAIR MRI in sporadic Creutzfeldt–Jakob disease. Neurology *56*, 282.

Zerr, I., Pocchiari, M., Collins, S., Brandel, J.P., de Pedro Cuesta, J., Knight, R.S., Bernheimer, H., Cardone, F., Delasnerie-Laupretre, N., Cuadrado Conales, N., *et al* (2000). Analysis of EEG and CSF 14-3-3 proteins as aids to the diagnosis of Creutzfeldt–Jakob disease. Neurology *55*, 811–815.

Zou, W.Q., Puoti, G., Xiao, X., Yuan, J., Qing, L., Cali, I., Shimoji, M., Langeveld, J.P., Castellani, R., Notari, S., *et al.* (2010). Variably protease-sensitive prionopathy: a new sporadic disease of the prion protein. Ann. Neurol. *68*, 162–172.

Immunological Strategies for the Prevention and Treatment of Prion Diseases

8

Keiji Uchiyama and Suehiro Sakaguchi

Abstract

Prion diseases, which include Creutzfeldt–Jakob disease (CJD) in humans and bovine spongiform encephalopathy (BSE) in animals, are a group of incurable neurodegenerative disorders caused by proteinaceous infectious agents, the so-called prions. No preventative vaccines and therapeutics of prion diseases have been developed. Recent lines of evidence suggest that antibodies against prion protein might be beneficial for both preventing and treating prion disease. In this chapter, we first discuss the possibility that there might be many individuals who are latently infected with vCJD prions in human populations, and then introduce the so far reported immunological approaches for development of prion vaccines and immunotherapy against prion disease, including our recent work.

Introduction

Prion diseases or transmissible spongiform encephalopathies including Creutzfeldt–Jakob disease (CJD) in humans, and bovine spongiform encephalopathy (BSE) and scrapie in animals, are a group of incurable neurodegenerative disorders caused by proteinaceous infectious agents, the so-called prions (DeArmond and Prusiner, 1995; Prusiner, 1998). Epidemiological studies have indicated the presence of a species barrier for scrapie prions between sheep or goats and humans (Chatelain *et al.*, 1981; Brown *et al.*, 1987). However, BSE prions are believed to overcome a species barrier between humans and cattle to transmit to humans, causing variant CJD (vCJD) in humans (Bruce *et al.*, 1997; Hill *et al.*, 1997). Three cases of vCJD considered to be transmitted via blood transfusion have been reported (Llewelyn *et al.*, 2004; Peden *et al.*, 2004; Wroe *et al.*, 2006), giving rise to the possibility of a vCJD epidemic in human population via contaminated blood. However, no preventative vaccines and therapeutics for prion diseases have yet been developed.

Prions are composed of the abnormally folded, relatively proteinase K-resistant, amyloidogenic isoform of prion protein, termed PrP^{Sc} (Prusiner, 1998). PrP^{Sc} is generated by conformational conversion of the normal cellular isoform of PrP, PrP^{C}, a glycosylphosphatidylinositol-anchored membrane glycoprotein abundantly expressed in neurons (Prusiner, 1998). Prions (or PrP^{Sc}) interact with PrP^{C} and induce changes in the protein conformation of the interacting PrP^{C} into that of PrP^{Sc} (Prusiner, 1998). The constitutive conversion of PrP^{C} into PrP^{Sc} leads to the propagation of prions, or the accumulation of PrP^{Sc} in the central nervous system (CNS). We and others previously demonstrated that mice devoid of PrP^{C}, in which the conversion of PrP^{C} into PrP^{Sc} cannot occur due to the absence of PrP^{C}, are resistant to prion diseases (Bueler *et al.*, 1993; Prusiner *et al.*, 1993; Manson *et al.*, 1994; Sakaguchi *et al.*, 1995).

PrP is a plausible target molecule for prion vaccines. Indeed, *in vitro* mixing of hamster-adapted scrapie prions with polyclonal antibodies (Abs) against PrP, α-PrP27–30, was reported to reduce the infectivity of the prions by a factor of 100 in animals (Gabizon *et al.*, 1988). Transgenic mice expressing a 6H4 anti-PrP monoclonal antibody (mAb) were also shown to be highly resistant to the disease even after intraperitoneal inoculation with mouse-adapted scrapie RML prions (Heppner *et al.*, 2001). It was also demonstrated that passive immunization with anti-PrP mAbs, ICSM 18 or 35, protected mice from the peripheral infection of RML prions (White *et al.*, 2003). However, PrP is a host protein expressed in various tissues, indicating that PrP is immunologically tolerated by the host (Oesch *et al.*, 1985). Therefore, overcoming immune tolerance to PrP is a crucial step for development of prion vaccines.

Anti-PrP Abs may be beneficial for treating prion diseases. Anti-PrP Fab fragments, termed D13, D18, R1, R2, E123, E149 and R72, and an anti-PrP 6H4 mAb were shown to reduce PrP^{Sc} levels in chronically infected N2a cells and eventually cure them (Enari *et al.*, 2001; Peretz *et al.*, 2001). However, Abs are macromolecules and are therefore unable to pass the blood–brain barrier (BBB). Indeed, it was shown that the intraperitoneally administrated anti-PrP ICSM 18 and 35 mAbs protected mice from the peripheral infection of RML prions, but had no effects on prions directly introduced into the brains of mice due to the inability of the Abs to cross the BBB (White *et al.*, 2003).

In this chapter, we first discuss the possibility of an epidemic of prion diseases in human populations, and then introduce the so far reported strategies to overcome immune tolerance to PrP and to deliver anti-PrP Abs into the brain, including our recent work.

Latent infection of prion disease in human populations

Prion diseases are rare in humans, with an incidence of about 1:1,000,000 worldwide (DeArmond and Prusiner, 1995). They manifest as sporadic, inherited and infectious disorders. Most of the cases, accounting for 85%–90%, are sporadic CJD of unknown aetiologies (DeArmond and Prusiner, 1995). About 10% of the cases are inherited, including those of familial CJD, Gerstmann–Sträussler–Scheinker syndrome, and fatal familial insomnia (DeArmond and Prusiner, 1995). These inherited diseases are aetiologically linked to specific mutations of *Prnp*, the gene for PrP (DeArmond and Prusiner, 1995). The infectious types of the disease, including vCJD, iatrogenic CJD and kuru, account for only several per cent of the cases (DeArmond and Prusiner, 1995). Kuru is a disease which emerged due to ritualistic cannibalism in Papua New Guinea (Gajdusek, 1977).

Patients affected with prion diseases usually die within a year after abnormal neurological symptoms are developed (Prusiner, 1996). However, prion diseases require a very long incubation period until neurological symptoms become evident (Prusiner, 1996), meaning that the patients are latently infected for long period, incubating prions in their bodies. Pre-symptomatic diagnosis of the disease is currently impossible. Therefore, these latently infected people become the sources of secondary transmission of the disease in human populations. Indeed, many cases of the transmission from the latently infected people have been so far reported. These transmissions accidentally happened via medical treatments and procedures, such as intracerebral insertion of electroencephalogram electrodes, which had been used for the latently infected people, and treatments with growth hormone preparations, dura matter or corneal grafts taken from the latently infected people (Duffy *et al.*, 1974; Bernoulli *et al.*, 1977; Koch *et al.*, 1985; Thadani *et al.*, 1988).

vCJD was first recognized in the UK in 1996 (Wells *et al.*, 1987). Since then, 176 cases of vCJD have been reported in the UK as of 2 July 2012 (http://www.cjd.ed.ac.uk/figures.htm) and a much lesser number of cases in other countries. In contrast to classical type CJD prions, vCJD prions can be present in blood in a considerably high titre. Unfortunately, three cases of vCJD considered to be transmitted via blood transfusion have been reported (Llewelyn *et al.*, 2004; Peden *et al.*, 2004; Wroe *et al.*, 2006). One case was diagnosed as vCJD by autopsy and the others developed the disease several years after blood transfusion from donors who eventually succumbed to vCJD. Blood transfusion is a common medical treatment. However, there are no appropriate measures to remove prions from the blood. It is therefore possible that vCJD could spread in human populations via medical treatments and procedures using the blood donated from the latently infected people with vCJD. Development of sensitive diagnostic measures that enable identification of latently infected people is urgently awaited.

Human *Prnp* is polymorphic at codon 129, coding methionine (M) or valine (V). This polymorphism is tightly linked to susceptibility to prion disease in humans (Cervenakova *et al.*, 1998; Brandel *et al.*, 2003; Wadsworth *et al.*, 2004). People with MM homozygous alleles are thought to be the most susceptible; those with MV and VV alleles are intermediate and resistant, respectively. All cases of vCJD transmitted from BSE, reported till date, are MM homozygous (Bishop *et al.*, 2006). No MV or VV cases have been identified (Bishop *et al.*, 2006). However, among the cases of blood transfusion-related vCJD, one case was heterozygous at codon 129 (Peden *et al.*, 2004), suggesting that vCJD prions could be transmissible to humans with any genotypes of *Prnp*. Indeed, vCJD was transmitted to mice expressing human PrP with MM, MV or VV alleles (Bishop *et al.*, 2006). It is therefore possible that people with not only MM but also MV or VV alleles could be latently infected with vCJD prions. If so, there might be many individuals who are latently infected with vCJD prions without any clinical symptoms in human populations. Hilton *et al.* reported a much greater incidence of the disease than that so far reported for conventional human prion diseases. They found that 3 out of 12,674 surgically removed appendectomy or tonsillectomy specimens were positive for staining of PrP^{Sc} (Hilton *et al.*, 2004).

More than 160,000 cases of BSE were reported in cattle in the UK. However, cases of BSE have dramatically decreased in number these days as a result of a ban on the use of meat and bonemeal ingredients in animal feed (Collee *et al.*, 2006), indicating that current risk of the transmission of BSE to humans has been markedly reduced. Indeed, the number of vCJD cases is gradually decreasing in the UK (http://www.cjd.ed.ac.uk/figures.htm). However, new cases of BSE are still reported in the UK and other countries. Therefore, constant survey of the disease is still necessary to monitor any outbreaks of BSE in the future. In North America, chronic wasting disease (CWD), another type of animal prion disease, is spreading within captive and free-ranging mule deer and elk populations, raising a new health concern about transmission of CWD to humans (Williams, 2005). Epidemiological studies indicated that risk, if any, of CWD transmission to humans is low (Belay *et al.*, 2001, 2004). Mice expressing transgenic human PrP were susceptible to human prions but not to elk CWD prions (Kong *et al.*, 2005). These results indicate the presence of a substantial species barrier for the transmission of CWD from deer to humans. However, the possibility of transmission of CWD to humans cannot be completely excluded at present. Therefore, CWD is also required to be constantly surveyed.

Possible anti-prion mechanisms of anti-PrP Abs

The mechanisms of how anti-PrP Abs inhibit the conformational conversion of PrP^{C} into PrP^{Sc} and exert anti-prion activities remain to be clarified. Several mechanisms have been proposed. The first step of the conversion is supposed to be an interaction between PrP^{C} and PrP^{Sc} probably taking place on the cell surface, particularly on lipid rafts, and/or along the endocytotic pathway (Harris, 1999). First, anti-PrP Abs might interfere with the interaction. 3F4 and 13A5 mAbs, which recognize residues 109–112 and 138–165, respectively, and the polyclonal Ab against residues 219–232 were shown to disturb the interaction and inhibit the conversion in a cell-free system (Kaneko *et al.*, 1995; Horiuchi and Caughey, 1999). Second, anti-PrP Abs might interfere with the interaction of the so-called co-factor(s), which is supposed to play an important role in the conversion, with either PrP^{C} or PrP^{Sc}, or both. Third, the subcellular localization of PrP^{C} might be altered by anti-PrP Abs. 31C6, 110, 44B1, and 72 anti-PrP mAbs were shown to disturb PrP^{C} internalization and reduce PrP^{Sc} levels in infected cells (Kim *et al.*, 2004). The 31C6 and 110 mAbs react with residues 143–149 and the PHGGGWG sequence in the octapeptide repeat region, respectively, and 44B1 and 72 mAbs recognize discontinuous epitopes (Kim *et al.*, 2004). Fourth, anti-PrP Abs might reduce PrP^{C} levels. Anti-PrP mAbs, SAF34 and SAF61, which react with the octapeptide repeat region and residues 144–152, respectively, were shown to accelerate degradation of PrP^{C} in cells (Perrier *et al.*, 2004).

Strategies to overcome immune tolerance to PrP

Molecular mimicry-based prion vaccines: heterologous PrP vaccines

Sigurdsson *et al.* reported that subcutaneous immunization with autologous recombinant mouse PrP induced anti-PrP auto-Abs in CD-1 mice and slightly retarded onset of the disease in mice subsequently inoculated with a mouse-adapted 139A prion (Sigurdsson *et al.*, 2002). However, other investigators showed that mice were tolerant to autologous recombinant mouse PrP, producing no anti-PrP auto-Abs (Polymenidou *et al.*, 2004). We also failed to detect anti-PrP auto-Abs in mice intraperitoneally immunized with autologous mouse recombinant PrP (Ishibashi *et al.*, 2006). No prophylactic effects of the immunization against the Fukuoka-1 mouse prion were observed in mice (Ishibashi *et al.*, 2006). These results indicate that autologous PrP might not be immunogenic enough to overcome tolerance.

Molecular mimicry between microbial and host antigens is a well-known hypothetical mechanism for production of auto-Abs and/or auto-reactive T cells due to similar amino acid sequences shared between both antigens (Ang *et al.*, 2004; Behar and Porcelli, 1995). We therefore hypothesized that heterologous PrPs, which are derived from a different species to the host, could overcome the immune tolerance to elicit Ab responses against the host PrP, because they are similar but not identical to the host PrP in amino acid sequence. We showed that heterologous sheep and bovine recombinant PrPs, but not autologous mouse recombinant PrP, were highly immunogenic and able to disrupt immune tolerance to the host PrP in BALB/c mice, inducing anti-PrP auto-Abs (Ishibashi *et al.*, 2007). No autoimmune-related symptoms such as arthritis or abnormal behaviour were detected in the immunized mice. Notably, prophylactic effects of the immunization with sheep or bovine recombinant PrP on infection with the Fukuoka-1 prion were observed in mice (Ishibashi *et al.*, 2007). Non-immunized BALB/c mice developed the disease at 291 ± 10 days post inoculation (dpi). However, BALB/c mice immunized with recombinant bovine PrP showed delayed onsets by 31 days. Recombinant sheep PrP exhibited variable effects against the prion in BALB/c mice. About 70% of the immunized mice developed the disease with prolonged onsets. These results indicate that immunization with heterologous recombinant PrPs might be protective against prions via inducing anti-PrP auto-Abs.

Molecular mimicry-based prion vaccines: non-PrP vaccines

Prion vaccines based on PrP molecules may risk causing adverse effects in the immunized host. First, it was shown that certain anti-PrP mAbs against residues 95–105 or the N-terminal octapeptide repeat region were neurotoxic in mice, causing neuronal cell death (Solforosi *et al.*, 2004; Lefebvre-Roque *et al.*, 2007). Thus, PrP vaccines might elicit such neurotoxic anti-PrP Abs in the immunized host. Second, since PrP is a host protein, PrP vaccines might cause autoimmune responses in the immunized host. Third, most importantly, it was reported that recombinant PrP was converted into infectious PrP *in vitro* by incubating with ubiquitous molecules, RNAs and lipids, and being subjected to protein misfolding cyclic amplification (Wang *et al.*, 2010). Therefore, immunizing PrP molecules in PrP vaccines may be converted into infectious PrPs or prions in the immunized host. These indicate that molecules other than PrP may be plausible as prion vaccines. We demonstrated that heterologous sheep and bovine recombinant PrPs function as antigen mimicking molecules to host mouse PrP, are highly immunogenic in BALB/c mice, efficiently inducing anti-PrP auto-Abs, and that immunization with the heterologous PrPs significantly prolonged incubation times in mice inoculated with the Fukuoka-1 prion (Ishibashi *et al.*, 2007). This suggested that certain molecules, if able to antigenically mimic anti-prion epitopes, could behave as prion vaccines.

We recently showed that the recombinant bacterial enzyme succinylarginine dihydrolase from *Escherichia coli* [hereafter referred to as SADH(*E*)] or *Salmonella enterica* subspecies *enterica* serovar Paratyphi A strain [hereafter referred to as SADH(*S*)] were highly immunogenic in mice, eliciting Ab responses against not only themselves but also the 6H4 anti-prion epitope (Fig. 8.1A) (Ishibashi *et al.*, 2011). SADH(*E*) and SADH(*S*) were searched for as molecules carrying a peptide sequence mimicking the 6H4 anti-prion epitope, which corresponds to residues 144–152 of mouse PrP, using Basic Local Alignment Search Tool in the National Center for Biotechnology Information website (http://blast.ncbi.nlm.nih.gov/Blast.cgi) (Fig. 8.1B). SADHs are a much conserved molecule involved in the arginine succinyltransferase pathway in *Escherichia coli* and related bacteria including *Salmonella enterica* subspecies (Cunin *et al.*, 1986). We then showed that anti-prion activities were induced in mice by immunization with SADH(*E*) and SADH(*S*). Anti-SADH(*E*) and anti-SADH(*S*) sera were able to prevent PrP^{Sc} formation in N2a cells that were persistently infected with Fukuoka-1 prion (Fig. 8.2) (Ishibashi *et al.*, 2011). We finally showed that immunization with the recombinant SADH(*E*) and SADH(*S*) significantly prolonged survival times

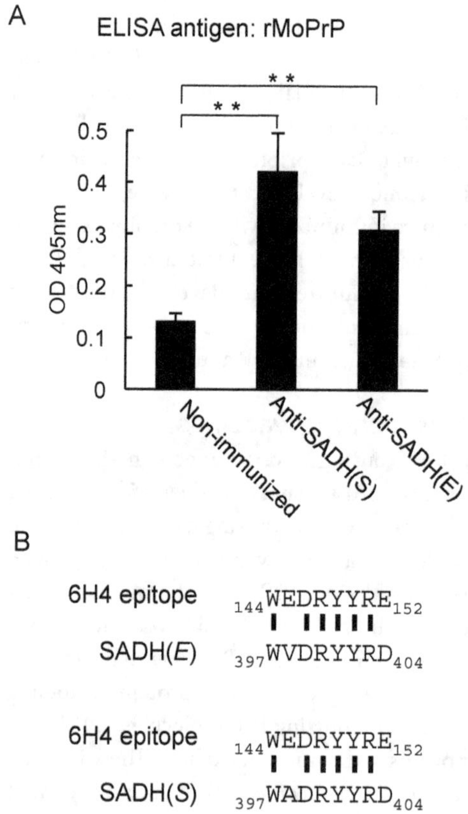

Figure 8.1 SADH recombinant proteins induce anti-PrP auto-Abs able to recognize the host PrP. (A) Anti-SADH(*E*) and anti-SADH(*S*) sera recognize full-length recombinant mouse PrP (rMoPrP). Anti-SADH(*E*) (n=10) and anti-SADH(*S*) sera (n=9) were diluted 1:20 and subjected to an ELISA against purified rMoPrP. Higher OD_{405} values were detected with anti-SADH(*E*) and anti-SADH(*S*) sera than those of control sera (n=9) from non-immunized mice. (B) Comparison of the amino acid sequence of the 6H4 anti-prion epitope to the mimic sequences of SADH(*E*) and SADH(*S*). Arabic numbers represent the codon number of each indicated amino acid in each molecule. (With permission from Elsevier; see Ishibashi *et al.* 2011.)

in mice subsequently infected with Fukuoka-1 prion (Table 8.1) (Ishibashi *et al.*, 2011). BALB/c mice were intraperitoneally immunized with purified recombinant SADH(*E*) and SADH(*S*) proteins five times at 2-week intervals and thereafter intraperitoneally inoculated with Fukuoka-1 prion 1 week after the final immunization. Non-immunized mice eventually died of the disease 298 ± 28 dpi. However, mice immunized with SADH(*E*) and SADH(*S*) recombinant proteins displayed significantly extended survival times by 31 (P=0.0284, logrank test) and 23 days (P=0.0384), respectively. These results indicate that the 6H4 mimicking sequence in SADH(*E*) and SADH(*S*) recombinant proteins was immunogenic in mice, eliciting anti-prion Abs that recognize the 6H4 epitope and thereby exerting anti-prion activities. However, the prophylactic effects of the immunization with SADH(*E*) and SADH(*S*) recombinant proteins were slight. This might be attributable to low titres of the anti-PrP auto-Abs produced in the immunized mice. Thus, enhancement of the antigenicity of the mimic sequence in SADH(*E*) and SADH(*S*) recombinant proteins is required to induce much higher titres of anti-prion Abs, enough to effectively prevent prion infection. Alternatively, it might be interesting to search for other non-mammalian molecules carrying an anti-prion epitope-mimic sequence(s) with stronger antigenic properties.

Mucosal vaccinations of PrP

Mucosal vaccines are able to stimulate local and systemic immune responses by inducing not only secretory IgA at mucosal surfaces but also serum IgG (Giudice and Campbell,

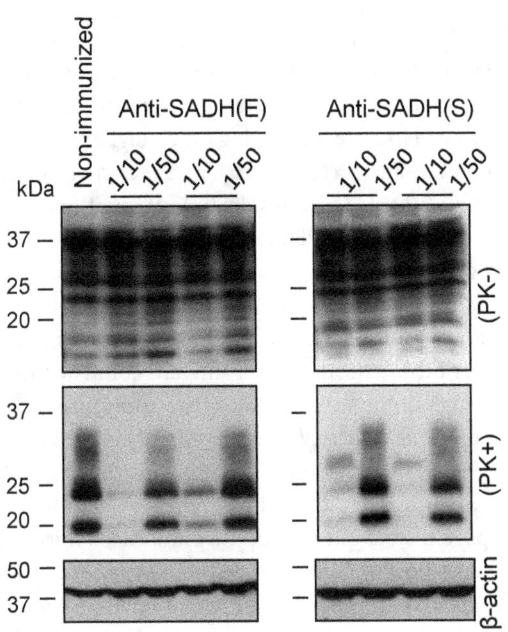

Figure 8.2 *In vitro* anti-prion activity of anti-SADH(*E*) and anti-SADH(*S*) sera. Anti-SADH(*E*) and anti-SADH(*S*) sera reduce PrP^{Sc} levels in N2a cells persistently infected with Fukuoka-1 prions. The cells were incubated for 2 days with anti-SADH(*E*) (n=2) and anti-SADH(*S*) sera (n=2) at indicated dilutions and then subjected to Western blotting. Compared wth non-immunized sera, anti-SADH(*E*) or anti-SADH(*S*) sera reduced PrP^{Sc} levels in the cells in a dose-dependent manner without affecting the level of total PrP. (With permission from Elsevier; see Ishibashi *et al.* 2011.)

Table 8.1 Prophylactic effects of immunization with SADHs on prion infection

Immunogen	Survival times (mean±SD, days)	No. of mice (diseased/total)	*P* values (logrank test)
–	298±28	10/10	
SADH(*E*)	329±15	5/5	0.0284
SADH(*S*)	321±15	5/5	0.0384

2006). We investigated the mucosal immunogenicity of PrP in mice following fusion with the heat-labile enterotoxin subunit B (LTB) of *Escherichia coli*. To do this, we generated LTB-moPrP120–231 and LTB–boPrP132–242 fusion proteins, in which the C-terminal residues 120–231 and 132–242 of mouse and bovine PrPs, respectively, were fused to the C-terminus of LTB (Yamanaka *et al.*, 2006). Co-immunization of non-fused moPrP120–231 with recombinant mutant non-toxic enterotoxin into nasal cavities of BALB/c mice induced no IgG response (Yamanaka *et al.*, 2006). However, LTB–moPrP120–231 fusion protein elicited significantly higher Ab responses in mice (Yamanaka *et al.*, 2006), indicating that fusion with LTB could enhance the mucosal immunogenicity of PrP. We also immunized mice with LTB–boPrP132–242 fusion protein as well as non-fusion boPrP132–242.

BoPrP132–242 itself moderately elicited an IgG response in mice probably due to its antigenic heterogeneity (Yamanaka *et al.*, 2006). However, no IgA response could be detected in the mice (Yamanaka *et al.*, 2006). In contrast, the mucosal immunogenicity of LTB-boPrP132–242 was enhanced in mice, producing much higher titres of anti-boPrP IgG and anti-boPrP IgA in serum (Yamanaka *et al.*, 2006). The specific IgA was also secreted in the intestines (Yamanaka *et al.*, 2006). These results indicate that fusion with LTB could augment the mucosal immunogenicity of PrP in mice.

Live-attenuated pathogenic *Salmonella* is an efficient mucosal delivery vector of antigens (De Magistris, 2006). *Salmonella* is commensal enteric bacteria. It is therefore conceivable that vaccinating *Salmonella* could continuously produce antigens in large amounts in the gut for a considerably long period and the produced antigens could be thus efficiently taken into the epithelium, eliciting high immune responses in the immunized mice. Goñi *et al.* (2005) produced an attenuated *Salmonella typhimurium* LVR01 LPS vaccine strain expressing mouse PrP fused with non-toxic fragment C of tetanus toxin and orally immunized CD-1 mice with it. The orally immunized mice elicited higher IgG and IgA responses against PrP (Goni *et al.*, 2005). Importantly, about 30% of the immunized mice were alive without any clinical signs up to 500 days post oral infection with 139A mouse prion (Goni *et al.*, 2005). The authors also subsequently showed that all of the mice producing high titres of anti-PrP IgG and IgA were completely resistant to the prion, being free of any disease-specific symptoms up to at least 400 days post oral infection (Goni *et al.*, 2008). In contrast, no significant extension of the survival times was observed in mice producing lower titres of anti-PrP Abs (Goni *et al.*, 2008). These results indicate that the *Salmonella* delivery system for PrP fused with the tetanus toxin fragment C could stimulate protective immunity against prion disease.

Dimeric PrP vaccine

Dimeric recombinant mouse PrP, which consists of a tandem duplication of mouse PrP with a human or hamster-derived 3F4 epitope replaced at the corresponding region, was shown to be immunogenic when subcutaneously immunized, eliciting anti-PrP auto-Abs in C57BL/6 mice (Gilch *et al.*, 2003). Monomeric mouse PrP with the 3F4 epitope was also immunogenic in mice (Gilch *et al.*, 2003). However, the Ab repertoire induced by dimeric and monomeric PrPs were different (Gilch *et al.*, 2003). Dimeric but not monomeric PrP produced Abs, which effectively inhibited PrP^{Sc} formation in persistently infected N2a cells (Gilch *et al.*, 2003), suggesting that dimeric PrP might stimulate protective immunity against prions. These PrPs contain a human or hamster-derived 3F4 epitope (Gilch *et al.*, 2003). It is therefore possible that the dimeric PrP might acquire heterologous PrP-like immunogenicities in part, resulting in production of anti-PrP auto-Abs with anti-prion activity.

Virus-like particle (VLP) vaccinations of PrP

VLPs are formed by self-assembly of virus-encoded capsid proteins and are known to stimulate immune responses (Noad and Roy, 2003). Nikles *et al.* generated murine leukaemia retrovirus-derived VLP displaying the C-terminal 111 amino acids of PrP (PrP111) fused to the transmembrane domain of platelet-derived growth factor receptor and subsequently immunized C57BL/6 mice with PrP111-VLP (Nikles *et al.*, 2005). PrP111-VLP induced anti-PrP auto-Abs, which can recognize the native form of PrP^{C} expressed on the cell surface (Nikles *et al.*, 2005), suggesting that PrP111-VPL could be effective as a prion vaccine.

Handisurya *et al.* produced bovine papillomavirus type 1-derived VPL displaying the nine amino acid epitope DWEDRYYRE of the rat PrP that was inserted into the L1 major capsid protein, and immunized rabbits and rats with it (Handisurya *et al.*, 2007). The DWEDRYYRE-VPL was more immunogenic in rabbits than in rats (Handisurya *et al.*, 2007). Rabbit anti-PrP serum contained higher titres of Abs than the rat anti-PrP serum and exhibited inhibition of *de novo* synthesis of PrP^{Sc} in infected N2a cells (Handisurya *et al.*, 2007). The corresponding amino acid sequence of rabbit PrP is different from the rat PrP peptide by one amino acid (DYEDRYYRE, the underline indicates a different amino acid) (Handisurya *et al.*, 2007). It is therefore possible that the augmented immune response in rabbits against DWEDRYYRE-VPL might be due to antigenic heterogeneity of the displayed peptide.

Other vaccination strategies

Heat shock proteins (Hsps) exert a strong adjuvant effect when coupled to an antigen (Zugel and Kaufmann, 1999; Srivastava, 2002). Antigens coupled to Hsps are captured by antigen-presenting cells via specific cellular surface receptors and displayed on cell surface MHC class I, stimulating immune responses (Zugel and Kaufmann, 1999; Srivastava, 2002). PrP conjugated with DnaK, a member of the heat shock protein 70 family, was shown to induce anti-PrP auto-Abs in BALB/c mice (Koller *et al.*, 2002).

A peptide simultaneously conjugated to a non-immunogenic dendritic scaffold molecule with multiple copies is more immunogenic, compared to a non-conjugated peptide. Indeed, conjugation of eight copies of a peptide corresponding to residues 144–153 (helix 1) of human PrP induced high titres of IgG specific to the peptide in BALB/c mice (Arbel *et al.*, 2003). However, the PrP peptide used in this immunization was different from that of mice by one amino acid (Arbel *et al.*, 2003). It is therefore possible that, rather than fusion of multiple PrP peptides, antigenic heterogeneity of the peptide might contribute to the disruption of immune tolerance.

DNA vaccines stimulate immune responses via uptake by professional antigen-presenting dendritic cell (DCs), where a DNA-encoded antigen is expressed and presented for T cell recognition, and by non-DCs such as keratinocytes or myocytes, which express the encoded antigens and transfer them to DCs (Djilali-Saiah *et al.*, 2002; Leitner *et al.*, 2003; Rice *et al.*, 2008). It was shown that immunization with plasmid pCMV-PrPLII, encoding mouse PrP fused to the lysosomal integral membrane protein type II lysosome-targeting signal, into the anterior tibial muscle of 129/ola mice stimulated production of anti-PrP auto-Abs and extended the incubation times by about 73 days in mice after inoculation with a mouse-adapted BSE_1 prion (Fernandez-Borges *et al.*, 2006). In contrast, pCMP-PrP which encodes mouse PrP alone, neither stimulated Ab responses in 129/ola mice nor protected the mice from the disease (Fernandez-Borges *et al.*, 2006). These results indicate that DNA vaccines encoding PrP fused with the lysosomal integral membrane protein type might be useful as a prion vaccine.

Pattern recognition Toll-like receptors recognize pathogenic organism-derived pathogen-associated molecular patterns, including unmethylated CpG oligodeoxynucleotides, and stimulate innate and ultimately acquired immune responses (O'Hagan *et al.*, 2001). It was reported that administration of CpG alone extended survival times in mice inoculated with the RML prions (Sethi *et al.*, 2002). The control mice developed the disease at 181 dpi (Sethi *et al.*, 2002). However, mice intraperitoneally administered with CpG 7 hours post-inoculation and thereafter daily for 20 days were free of the disease more than 330 dpi. It was thereafter shown that such repeated administration of CpG caused suppression of follicular dendritic cells (FDCs), which important for prion propagation before invasion of the CNS (Heikenwalder *et al.*, 2004). It is therefore possible that the prophylactic effects of CpG are due to suppression of FDCs, caused by its repeated administration.

Treatment strategies of prion diseases

Intraventricular delivery of anti-PrP Abs

We directly delivered 3S9 anti-PrP mAb into the brains of mice via an intraventricular route. 3S9 mAb recognizes residues 141–161 and exerts anti-prion activity, reducing PrP^{Sc} levels in infected N2a cells with a 50% inhibitory concentration (IC_{50}) value of 0.6 nM (Miyamoto *et al.*, 2005; Sakaguchi *et al.*, 2009). The mAb was continuously delivered into the right ventricle of mice that had been intracerebrally inoculated with a mouse-adapted Fukuoka-1 prion using the ALLZET mini-osmotic pump model 2004 (DURECT Corporation) at a flow rate of 0.25 ± 0.05 μg/h for 28 days from 8 or 13 weeks post inoculation (wpi) (Sakaguchi *et al.*, 2009). No significant extension of incubation times and survival times was detected in the treated mice. These results suggest that, probably due to high molecular weight, the intraventricularly administered 3S9 mAb is unable to infiltrate brain regions where neurons mediating vital functions are infected by prions at concentrations that are sufficiently high for the Ab to exert therapeutic effects on the disease progression.

Song *et al.* reported marginal but significant therapeutic effects of anti-PrP Ab in animal prion disease models (Song *et al.*, 2008). They directly infused 31C6 anti-PrP mAb (0.7 nM IC_{50}) into the left ventricle of mice for 28 days from 60, 90 or 120 dpi with Chandler prions or Obihiro prions. Mice infected with Chandler prions survived the disease by approximately 10 days longer than control mice, regardless of the time points of treatment. On the other hand, the mice infected with Obihiro prions survived the disease by about 10 days longer than control mice, only when the treatment was started at 60 dpi, but not at 90 and 120 dpi. These results indicate that the anti-prion effect of 31C6 mAb were different for each of the prion strains. They used 31C6 mAb (0.5 μg/h) at twice the dose of 3S9 mAb we used (0.25 μg/h). It is therefore possible that a higher dose of the Ab might contribute to the extension of survival times in the mice. Alternatively, Fukuoka-1 prions might be more resistant to the anti-PrP Ab-mediated immunotherapy than Chandler or Obihiro prions.

Virus vector-mediated gene delivery of anti-PrP Abs

Wuertzer *et al.* (2008) used a recombinant adeno-associated vector type 2 (rAAV2) to deliver anti-PrP single chain Ab fragments (scFv) into the brain. They generated rAAV2 encoding anti-PrP scFv fragments, termed scFv3:3, scFv6:4, scFv6:6, and scFvD18, and bilaterally injected 9×10^9 expression units of them into the thalamus and striatum of mice 1 month before intraperitoneal inoculation with RML prions (Wuertzer *et al.*, 2008). Mice treated with control vector developed the disease at 199 ± 1 dpi. No significant prolongation in incubation times was detected in mice injected with rAAV-scFv6:4 and –scFv6:6. However, mice injected with rAAV-scFvs3:3 and -scFvD18 showed significantly extended incubation times by 23 and 51 days, respectively. PrP^{Sc} less accumulated in the brains of mice injected with rAAV-scFvsD18. The different anti-prion activities of the rAAV-scFvs were correlated to the binding affinity of the scFvs to recombinant PrP. These results indicate that the rAAV2 vector might be useful as a delivery vector for anti-PrP scFv fragments into the brain. However, regions expressing scFvs seemed restricted to sites where the rAAV2 scFvs were injected.

Brain-engraftable microglia-mediated *ex vivo* gene delivery of anti-PrP Ab

Microglial cells infiltrate and accumulate at the pathological lesions affected by prions (Kopacek *et al.*, 2000). Since an immortalized rat microglial cell line expressing the *ex vivo* transfected-lacZ gene was shown to be engrafted into the rat brain, surviving for at least the 3 weeks it was monitored (Sawada *et al.*, 1998), we investigated whether or not an immortalized murine microglial cell line could deliver anti-PrP Abs into the brain (Fujita *et al.*, 2011). We *ex vivo* transfected a vector encoding 3S9 anti-PrP scFv (3S9scFv) into the already established murine microglial Ra2 cell line (Kanzawa *et al.*, 2000), generating a Ra2 cell line permanently expressing 3S9scFv, designated 3S9scFv/GFP-Ra2 (Fig. 8.3A and B) (Fujita *et al.*, 2011). We directly injected 3S9scFv/GFP-Ra2 microglial cells into the right cerebrum at 1 week and into the left cerebrum at 3 weeks before inoculation of mouse-adapted Chandler prions into the right cerebrum (Fujita *et al.*, 2011). Survival times of the mice injected with 3S9scFv/GFP-Ra2 cells were marginally but still significantly elongated by 5 days, compared to those of the control mice ($P = 0.014$, logrank test) (Fig. 8.4). These results indicate that 3S9scFv delivered from the microglia was prophylactically effective against Chandler prions.

We then investigated the therapeutic effects of 3S9scFv/GFP-Ra2 cells against Chandler prions and 22L prions (Fujita *et al.*, 2011). 3S9scFv/GFP-Ra2 microglial cells were injected into the right cerebrum of mice 7 or 13 wpi with Chandler or 22L scrapie prions. No significant elongation of the survival times was observed in the mice infected with Chandler prions (Fig. 8.5). However, 3S9scFv/GFP-Ra2 cells were partially but still significantly effective against 22L prions when injected 7 wpi ($P = 0.035$, logrank test), extending survival times by 5 days, but not 13 wpi (Fig. 8.5). This is possibly because different strains of prions affect different brain regions and therefore the brain regions relevant to therapy might be somewhat different from one strain to another. Taken together, these results suggest that the brain-engraftable cell-mediated *ex vivo* gene transfer of anti-PrP Abs might be a possible immunotherapeutic approach against prion diseases.

The brain-engraftable cell-mediated *ex vivo* gene transfer of anti-PrP Abs into the brain may be advantageous as an immunotherapeutic approach against prion diseases over the direct intraventricular infusion or the virus vector-mediated gene transfer methods. The intraventricularly infused Abs and the scFvs delivered by virus vectors spread to only restricted regions that are close to sites where the Abs or virus vectors are injected, resulting in a limited effect on PrP^{Sc} formation and prion propagation in the brain. In contrast, the brain-engraftable cells have the potential to migrate to broad regions after engraftment into the brain. Therefore, more therapeutic effects may be expected with the brain-engraftable cell-mediated *ex vivo* gene transfer of anti-PrP Abs. Indeed, bone marrow-derived mesenchymal stem cells were reported to spread widely throughout the brains of mice that had been infected with prions, when engrafted or even intravenously injected into the hippocampus (Song *et al.*, 2009).

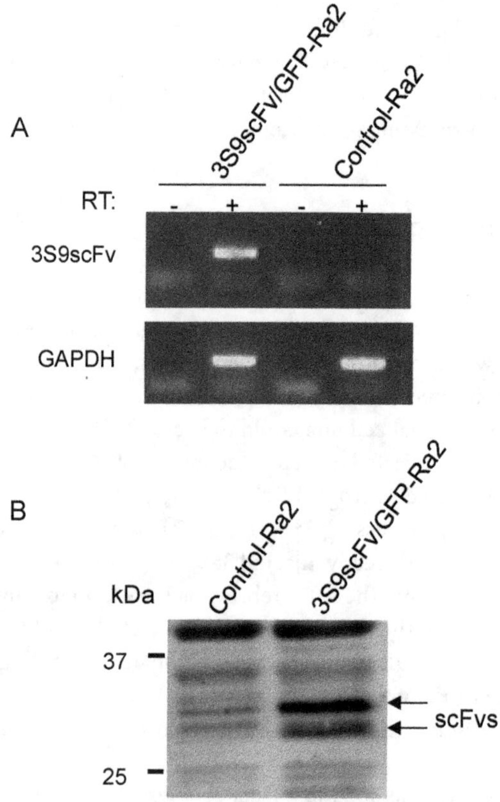

Figure 8.3 3S9scFv expressed by Ra2 cells. (A) Reverse transcriptase-polymerase chain reaction for 3S9scFv. 3S9scFv mRNA expression was detected in 3S9scFv/GFP-Ra2 cells but not in control-Ra2 cells. Glyceraldehyde-3-phosphate dehydrogenase (GAPDH) mRNA was similarly expressed in both cells. (B) Western blotting for 3S9scFv. Lysates from control-Ra2 and 3S9scFv/GFP-Ra2 cells were subjected to Western blotting. Glycosylated and unglycosylated 3S9scFvs were expressed in 3S9scFv/GFP-Ra2 cells. (With permission from Springer; see Fujita *et al.* 2011.)

Neurotoxic anti-PrP Abs

We detected no neuronal loss in the brains of mice administered with 3S9 mAb (Sakaguchi *et al.*, 2009). No neurotoxicity was also reported in mice injected with 31C6 mAbs (Song *et al.*, 2008). However, it was shown that certain anti-PrP Abs were neurotoxic. Lefebvre-Roque *et al.* reported that 4H11 anti-PrP mAb or its $F(ab')_2$ fragment induced extensive neuronal cell death and marked gliosis over the brain when administered daily for 2 weeks into the lateral ventricle of PrP-overexpressing Tg20 mice that had been infected with the 6PB1 mouse-adapted BSE prion (Lefebvre-Roque *et al.*, 2007). The neuronal loss was observed in regions close to the injected lateral ventricle as well as the occipital cortex, the hippocampus, the thalamus and the striatum. Other investigators also reported that anti-PrP mAbs D13 and P were toxic causing neuronal cell death even in normal mice when directly injected into the hippocampus or the cerebellar cortex (Solforosi *et al.*, 2004). The neurotoxic 4H11, D13, and P mAbs bind to epitopes located within the N-terminal part of PrP. 4H11 mAb binds to the OR region (residues 51–90), and D13 and P mAbs recognize epitopes within

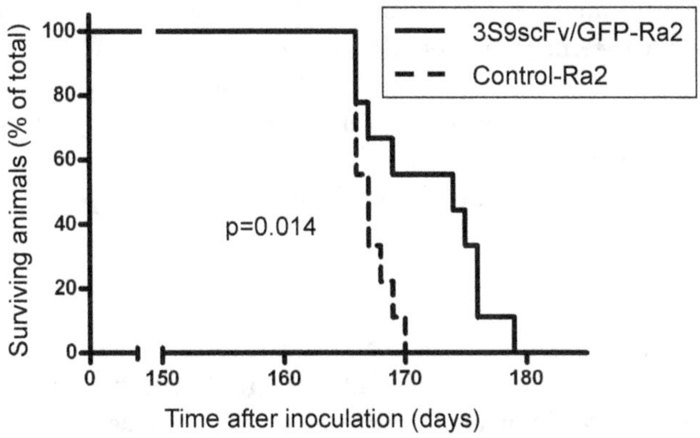

Figure 8.4 Survival curves of mice prophylactically injected with 3S9scFv/GFP-Ra2 cells. Mice (n=10) intracerebrally injected with 3S9scFv/GFP-Ra2 cells before infection with Chandler prions survived significantly longer than those (n=10) injected with control-Ra2 cells (p=0.014, logrank test). (With permission from Springer; see Fujita *et al*. 2011.)

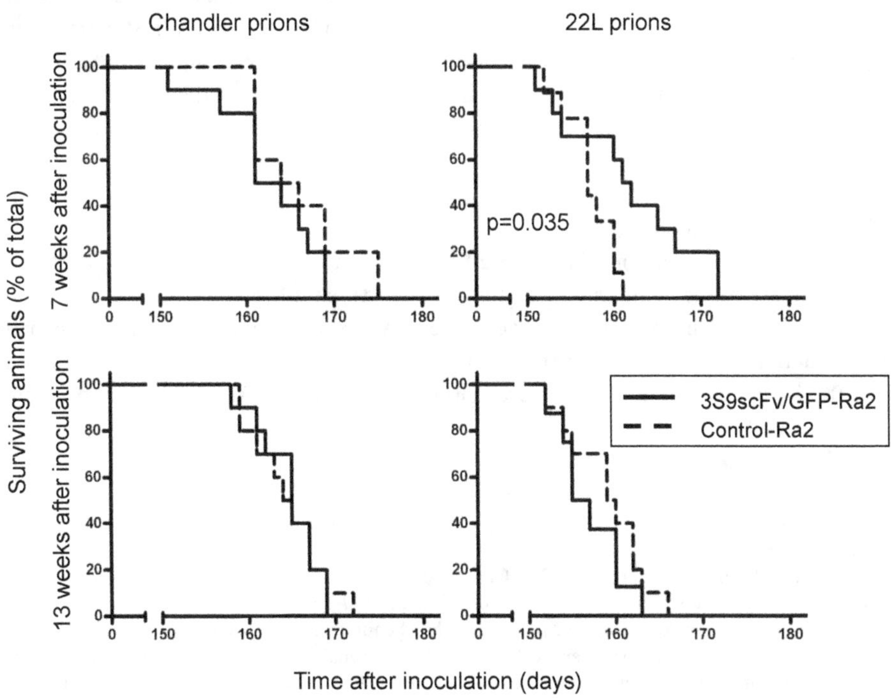

Figure 8.5 Survival curves of mice intracerebrally injected with 3S9scFv/GFP-Ra2 cells 7 and 12 weeks after infection with prions. No anti-prion effects of 3S9scFv/GFP-Ra2 cells were detected in mice infected with Chandler prions. In contrast, injection of 3S9scFv/GFP-Ra2 cells 7 weeks after infection with 22L prions significantly extended survival times in mice, compared to injection of control-Ra2 cells (P=0.035, Logrank test). No extension in survival times could be detected in 22L prion-infected mice when 3S9scFv/GFP-Ra2 cells were injected 13 weeks after infection commenced. Each group comprised nine or ten mice. (With permission from Springer; see Fujita *et al*. 2011.)

residues 95–105. In contrast, non-toxic 3S9 and 31C6 mAbs bind to epitopes within residues 141–161 of the C-terminal part of PrP. Thus, binding of anti-PrP antibodies to certain regions within the N-terminal part, such as the OR region or residues 95–105, might elicit a neurotoxic signal by either activating or adversely preventing the physiological function of PrP^C in neurons. Thus, an appropriate selection of non-neurotoxic anti-PrP mAbs, such as 31C6 and 3S9 mAbs, may be required for use in immunotherapy of prion diseases.

Perspectives

Pluripotent stem cells, including bone marrow stem cells, bone marrow-derived mesenchymal stem cells and induced pluripotent stem (iPS) cells, provide an attractive cell replacement therapy in central nervous disorders, including stroke and neurodegenerative diseases. This is because these cells have the potential to differentiate into neuronal or glial cells in the brain (Bobis *et al.*, 2006; Bae *et al.*, 2007; Chamberlain *et al.*, 2008; Wernig *et al.*, 2008; Bahat-Stroomza *et al.*, 2009; Dharmasaroja, 2009). Also, these stem cells could be available as cell sources for gene transfer to the brain. Indeed, the transplanted bone marrow cells, which were genetically engineered to express the green fluorescence protein, migrated into the brain parenchyma of many regions of the brain (Simard and Rivest, 2004; Wernig *et al.*, 2008). Wernig *et al.* (2008) also reported that fibroblast-derived iPS migrated into various brain regions and differentiated into glial and neuronal cells upon transplantation into the fetal mouse brain. It is thus worthwhile to investigate whether pluripotent stem cells expressing anti-PrP Abs or scFv fragments could be therapeutic against prion diseases.

Acknowledgements

This study was partly supported by a Grant-in-Aid from the BSE and other Prion Disease Control Project of the Ministry of Agriculture, Forestry and Fisheries of Japan, Grants-in-Aid from the Research Committee of Prion Disease and Slow Virus infection, the Ministry of Health, Labour and Welfare of Japan, and a Grant-in-Aid for TSE research (H23-Shokuhin-Ippan-005) and Research on Measures for Intractable Diseases from the Ministry of Health, Labour and Welfare of Japan.

References

Ang, C.W., Jacobs, B.C., and Laman, J.D. (2004). The Guillain–Barré syndrome: a true case of molecular mimicry. Trends Immunol. 25, 61–66.

Arbel, M., Lavie, V., and Solomon, B. (2003). Generation of antibodies against prion protein in wild-type mice via helix 1 peptide immunization. J. Neuroimmunol. *144*, 38–45.

Bae, J.S., Han, H.S., Youn, D.H., Carter, J.E., Modo, M., Schuchman, E.H., and Jin, H.K. (2007). Bone marrow-derived mesenchymal stem cells promote neuronal networks with functional synaptic transmission after transplantation into mice with neurodegeneration. Stem Cells 25, 1307–1316.

Bahat-Stroomza, M., Barhum, Y., Levy, Y.S., Karpov, O., Bulvik, S., Melamed, E., and Offen, D. (2009). Induction of adult human bone marrow mesenchymal stromal cells into functional astrocyte-like cells: potential for restorative treatment in Parkinson's disease. J. Mol. Neurosci. *39*, 199–210.

Behar, S.M., and Porcelli, S.A. (1995). Mechanisms of autoimmune disease induction. The role of the immune response to microbial pathogens. Arthritis Rheum. *38*, 458–476.

Belay, E.D., Gambetti, P., Schonberger, L.B., Parchi, P., Lyon, D.R., Capellari, S., McQuiston, J.H., Bradley, K., Dowdle, G., Crutcher, J.M., *et al.* (2001). Creutzfeldt–Jakob disease in unusually young patients who consumed venison. Arch. Neurol. *58*, 1673–1678.

Belay, E.D., Maddox, R.A., Williams, E.S., Miller, M.W., Gambetti, P., and Schonberger, L.B. (2004). Chronic wasting disease and potential transmission to humans. Emerg. Infect. Dis. *10*, 977–984.

Bernoulli, C., Siegfried, J., Baumgartner, G., Regli, F., Rabinowicz, T., Gajdusek, D.C., and Gibbs, C.J., Jr. (1977). Danger of accidental person-to-person transmission of Creutzfeldt–Jakob disease by surgery. Lancet *1*, 478–479.

Bishop, M.T., Hart, P., Aitchison, L., Baybutt, H.N., Plinston, C., Thomson, V., Tuzi, N.L., Head, M.W., Ironside, J.W., Will, R.G., *et al.* (2006). Predicting susceptibility and incubation time of human-to-human transmission of vCJD. Lancet Neurol. *5*, 393–398.

Bobis, S., Jarocha, D., and Majka, M. (2006). Mesenchymal stem cells: characteristics and clinical applications. Folia Histochem. Cytobiol. *44*, 215–230.

Brandel, J.P., Preece, M., Brown, P., Croes, E., Laplanche, J.L., Agid, Y., Will, R., and Alperovitch, A. (2003). Distribution of codon 129 genotype in human growth hormone-treated CJD patients in France and the UK. Lancet *362*, 128–130.

Brown, P., Cathala, F., Raubertas, R.F., Gajdusek, D.C., and Castaigne, P. (1987). The epidemiology of Creutzfeldt–Jakob disease: conclusion of a 15-year investigation in France and review of the world literature. Neurology *37*, 895–904.

Bruce, M.E., Will, R.G., Ironside, J.W., McConnell, I., Drummond, D., Suttie, A., McCardle, L., Chree, A., Hope, J., Birkett, C., *et al.* (1997). Transmissions to mice indicate that 'new variant' CJD is caused by the BSE agent. Nature *389*, 498–501.

Bueler, H., Aguzzi, A., Sailer, A., Greiner, R.A., Autenried, P., Aguet, M., and Weissmann, C. (1993). Mice devoid of PrP are resistant to scrapie. Cell *73*, 1339–1347.

Cervenakova, L., Goldfarb, L.G., Garruto, R., Lee, H.S., Gajdusek, D.C., and Brown, P. (1998). Phenotype–genotype studies in kuru: implications for new variant Creutzfeldt–Jakob disease. Proc. Natl. Acad. Sci. U.S.A. *95*, 13239–13241.

Chamberlain, S.J., Li, X.J., and Lalande, M. (2008). Induced pluripotent stem (iPS) cells as *in vitro* models of human neurogenetic disorders. Neurogenetics *9*, 227–235.

Chatelain, J., Cathala, F., Brown, P., Raharison, S., Court, L., and Gajdusek, D.C. (1981). Epidemiologic comparisons between Creutzfeldt–Jakob disease and scrapie in France during the 12-year period 1968–1979. J. Neurol. Sci. *51*, 329–337.

Collee, J.G., Bradley, R., and Liberski, P.P. (2006). Variant CJD (vCJD) and bovine spongiform encephalopathy (BSE): 10 and 20 years on: part 2. Folia Neuropathol. *44*, 102–110.

Cunin, R., Glansdorff, N., Pierard, A., and Stalon, V. (1986). Biosynthesis and metabolism of arginine in bacteria. Microbiol. Rev. *50*, 314–352.

De Magistris, M.T. (2006). Mucosal delivery of vaccine antigens and its advantages in pediatrics. Adv. Drug Deliv. Rev. *58*, 52–67.

DeArmond, S.J., and Prusiner, S.B. (1995). Etiology and pathogenesis of prion diseases. Am J. Pathol. *146*, 785–811.

Dharmasaroja, P. (2009). Bone marrow-derived mesenchymal stem cells for the treatment of ischemic stroke. J. Clin. Neurosci. *16*, 12–20.

Djilali-Saiah, I., Lapierre, P., Vittozi, S., and Alvarez, F. (2002). DNA vaccination breaks tolerance for a neo-self antigen in liver: a transgenic murine model of autoimmune hepatitis. J. Immunol. *169*, 4889–4896.

Duffy, P., Wolf, J., Collins, G., DeVoe, A.G., Streeten, B., and Cowen, D. (1974). Letter: Possible person-to-person transmission of Creutzfeldt–Jakob disease. N. Engl. J. Med. *290*, 692–693.

Enari, M., Flechsig, E., and Weissmann, C. (2001). Scrapie prion protein accumulation by scrapie-infected neuroblastoma cells abrogated by exposure to a prion protein antibody. Proc. Natl. Acad. Sci. U.S.A. *98*, 9295–9299.

Fernandez-Borges, N., Brun, A., Whitton, J.L., Parra, B., Diaz-San Segundo, F., Salguero, F.J., Torres, J.M., and Rodriguez, F. (2006). DNA vaccination can break immunological tolerance to PrP in wild-type mice and attenuates prion disease after intracerebral challenge. J. Virol. *80*, 9970–9976.

Fujita, K., Yamaguchi, Y., Mori, T., Muramatsu, N., Miyamoto, T., Yano, M., Miyata, H., Ootsuyama, A., Sawada, M., Matsuda, H., *et al.* (2011). Effects of a brain-engraftable microglial cell line expressing anti-prion scFv antibodies on survival times of mice infected with scrapie prions. Cell. Mol. Neurobiol. *31*, 999–1008.

Gabizon, R., McKinley, M.P., Groth, D., and Prusiner, S.B. (1988). Immunoaffinity purification and neutralization of scrapie prion infectivity. Proc. Natl. Acad. Sci. U.S.A. *85*, 6617–6621.

Gajdusek, D.C. (1977). Unconventional viruses and the origin and disappearance of kuru. Science *197*, 943–960.

Gilch, S., Wopfner, F., Renner-Muller, I., Kremmer, E., Bauer, C., Wolf, E., Brem, G., Groschup, M.H., and Schatzl, H.M. (2003). Polyclonal anti-PrP auto-antibodies induced with dimeric PrP interfere efficiently with PrPSc propagation in prion-infected cells. J. Biol. Chem. *278*, 18524–18531.

Giudice, E.L., and Campbell, J.D. (2006). Needle-free vaccine delivery. Adv. Drug Deliv. Rev. *58*, 68–89.

Goni, F., Knudsen, E., Schreiber, F., Scholtzova, H., Pankiewicz, J., Carp, R., Meeker, H.C., Rubenstein, R., Brown, D.R., Sy, M.S., *et al.* (2005). Mucosal vaccination delays or prevents prion infection via an oral route. Neuroscience *133*, 413–421.

Goni, F., Prelli, F., Schreiber, F., Scholtzova, H., Chung, E., Kascsak, R., Brown, D.R., Sigurdsson, E.M., Chabalgoity, J.A., and Wisniewski, T. (2008). High titers of mucosal and systemic anti-PrP antibodies abrogate oral prion infection in mucosal-vaccinated mice. Neuroscience *153*, 679–686.

Handisurya, A., Gilch, S., Winter, D., Shafti-Keramat, S., Maurer, D., Schatzl, H.M., and Kirnbauer, R. (2007). Vaccination with prion peptide-displaying papillomavirus-like particles induces autoantibodies to normal prion protein that interfere with pathologic prion protein production in infected cells. FEBS J. *274*, 1747–1758.

Harris, D.A. (1999). Cellular biology of prion diseases. Clin. Microbiol. Rev. *12*, 429–444.

Heikenwalder, M., Polymenidou, M., Junt, T., Sigurdson, C., Wagner, H., Akira, S., Zinkernagel, R., and Aguzzi, A. (2004). Lymphoid follicle destruction and immunosuppression after repeated CpG oligodeoxynucleotide administration. Nat. Med. *10*, 187–192.

Heppner, F.L., Musahl, C., Arrighi, I., Klein, M.A., Rulicke, T., Oesch, B., Zinkernagel, R.M., Kalinke, U., and Aguzzi, A. (2001). Prevention of scrapie pathogenesis by transgenic expression of anti-prion protein antibodies. Science *294*, 178–182.

Hill, A.F., Desbruslais, M., Joiner, S., Sidle, K.C., Gowland, I., Collinge, J., Doey, L.J., and Lantos, P. (1997). The same prion strain causes vCJD and BSE. Nature *389*, 448–450, 526.

Hilton, D.A., Ghani, A.C., Conyers, L., Edwards, P., McCardle, L., Ritchie, D., Penney, M., Hegazy, D., and Ironside, J.W. (2004). Prevalence of lymphoreticular prion protein accumulation in UK tissue samples. J. Pathol. *203*, 733–739.

Horiuchi, M., and Caughey, B. (1999). Specific binding of normal prion protein to the scrapie form via a localized domain initiates its conversion to the protease-resistant state. EMBO J. *18*, 3193–3203.

Ishibashi, D., Yamanaka, H., Yamaguchi, N., Yoshikawa, D., Nakamura, R., Okimura, N., Yamaguchi, Y., Shigematsu, K., Katamine, S., and Sakaguchi, S. (2006). Immunization with recombinant bovine but not mouse prion protein delays the onset of disease in mice inoculated with a mouse-adapted prion. Vaccine *25*, 985–992.

Ishibashi, D., Yamanaka, H., Yamaguchi, N., Yoshikawa, D., Nakamura, R., Okimura, N., Yamaguchi, Y., Shigematsu, K., Katamine, S., and Sakaguchi, S. (2007). Immunization with recombinant bovine but not mouse prion protein delays the onset of disease in mice inoculated with a mouse-adapted prion. Vaccine *25*, 985–992.

Ishibashi, D., Yamanaka, H., Mori, T., Yamaguchi, N., Yamaguchi, Y., Nishida, N., and Sakaguchi, S. (2011). Antigenic mimicry-mediated anti-prion effects induced by bacterial enzyme succinylarginine dihydrolase in mice. Vaccine *29*, 9321–9328.

Kaneko, K., Peretz, D., Pan, K.M., Blochberger, T.C., Wille, H., Gabizon, R., Griffith, O.H., Cohen, F.E., Baldwin, M.A., and Prusiner, S.B. (1995). Prion protein (PrP) synthetic peptides induce cellular PrP to acquire properties of the scrapie isoform. Proc. Natl. Acad. Sci. U.S.A. *92*, 11160–11164.

Kanzawa, T., Sawada, M., Kato, K., Yamamoto, K., Mori, H., and Tanaka, R. (2000). Differentiated regulation of allo-antigen presentation by different types of murine microglial cell lines. J. Neurosci. Res. *62*, 383–388.

Kim, C.L., Karino, A., Ishiguro, N., Shinagawa, M., Sato, M., and Horiuchi, M. (2004). Cell-surface retention of PrPC by anti-PrP antibody prevents protease-resistant PrP formation. J. Gen. Virol. *85*, 3473–3482.

Koch, T.K., Berg, B.O., De Armond, S.J., and Gravina, R.F. (1985). Creutzfeldt–Jakob disease in a young adult with idiopathic hypopituitarism. Possible relation to the administration of cadaveric human growth hormone. N. Engl. J. Med. *313*, 731–733.

Koller, M.F., Grau, T., and Christen, P. (2002). Induction of antibodies against murine full-length prion protein in wild-type mice. J. Neuroimmunol. *132*, 113–116.

Kong, Q., Huang, S., Zou, W., Vanegas, D., Wang, M., Wu, D., Yuan, J., Zheng, M., Bai, H., Deng, H., *et al.* (2005). Chronic wasting disease of elk: transmissibility to humans examined by transgenic mouse models. J. Neurosci. *25*, 7944–7949.

Kopacek, J., Sakaguchi, S., Shigematsu, K., Nishida, N., Atarashi, R., Nakaoke, R., Moriuchi, R., Niwa, M., and Katamine, S. (2000). Up-regulation of the genes encoding lysosomal hydrolases, a perforin-like protein, and peroxidases in the brains of mice affected with an experimental prion disease. J. Virol. *74*, 411–417.

Lefebvre-Roque, M., Kremmer, E., Gilch, S., Zou, W.Q., Feraudet, C., Gilles, C.M., Sales, N., Grassi, J., Gambetti, P., Baron, T., *et al.* (2007). Toxic effects of intracerebral PrP antibody administration during the course of BSE infection in mice. Prion *1*, 198–206.

Leitner, W.W., Hwang, L.N., deVeer, M.J., Zhou, A., Silverman, R.H., Williams, B.R., Dubensky, T.W., Ying, H., and Restifo, N.P. (2003). Alphavirus-based DNA vaccine breaks immunological tolerance by activating innate antiviral pathways. Nat. Med. *9*, 33–39.

Llewelyn, C.A., Hewitt, P.E., Knight, R.S., Amar, K., Cousens, S., Mackenzie, J., and Will, R.G. (2004). Possible transmission of variant Creutzfeldt–Jakob disease by blood transfusion. Lancet *363*, 417–421.

Manson, J.C., Clarke, A.R., McBride, P.A., McConnell, I., and Hope, J. (1994). PrP gene dosage determines the timing but not the final intensity or distribution of lesions in scrapie pathology. Neurodegeneration *3*, 331–340.

Miyamoto, K., Nakamura, N., Aosasa, M., Nishida, N., Yokoyama, T., Horiuchi, H., Furusawa, S., and Matsuda, H. (2005). Inhibition of prion propagation in scrapie-infected mouse neuroblastoma cell lines using mouse monoclonal antibodies against prion protein. Biochem. Biophys. Res. Commun. *335*, 197–204.

Nikles, D., Bach, P., Boller, K., Merten, C.A., Montrasio, F., Heppner, F.L., Aguzzi, A., Cichutek, K., Kalinke, U., and Buchholz, C.J. (2005). Circumventing tolerance to the prion protein (PrP): vaccination with PrP-displaying retrovirus particles induces humoral immune responses against the native form of cellular PrP. J. Virol. *79*, 4033–4042.

Noad, R., and Roy, P. (2003). Virus-like particles as immunogens. Trends Microbiol. *11*, 438–444.

O'Hagan, D.T., MacKichan, M.L., and Singh, M. (2001). Recent developments in adjuvants for vaccines against infectious diseases. Biomol. Eng. *18*, 69–85.

Oesch, B., Westaway, D., Walchli, M., McKinley, M.P., Kent, S.B., Aebersold, R., Barry, R.A., Tempst, P., Teplow, D.B., Hood, L.E., *et al.* (1985). A cellular gene encodes scrapie PrP 27–30 protein. Cell *40*, 735–746.

Peden, A.H., Head, M.W., Ritchie, D.L., Bell, J.E., and Ironside, J.W. (2004). Preclinical vCJD after blood transfusion in a *PRNP* codon 129 heterozygous patient. Lancet *364*, 527–529.

Peretz, D., Williamson, R.A., Kaneko, K., Vergara, J., Leclerc, E., Schmitt-Ulms, G., Mehlhorn, I.R., Legname, G., Wormald, M.R., Rudd, P.M., *et al.* (2001). Antibodies inhibit prion propagation and clear cell cultures of prion infectivity. Nature *412*, 739–743.

Perrier, V., Solassol, J., Crozet, C., Frobert, Y., Mourton-Gilles, C., Grassi, J., and Lehmann, S. (2004). Anti-PrP antibodies block PrPSc replication in prion-infected cell cultures by accelerating PrPC degradation. J. Neurochem. *89*, 454–463.

Polymenidou, M., Heppner, F.L., Pellicioli, E.C., Urich, E., Miele, G., Braun, N., Wopfner, F., Schatzl, H.M., Becher, B., and Aguzzi, A. (2004). Humoral immune response to native eukaryotic prion protein correlates with anti-prion protection. Proc. Natl. Acad. Sci. U.S.A. *101*, 14670–14676.

Prusiner, S.B. (1996). Human prion diseases and neurodegeneration. Curr. Top. Microbiol. Immunol. *207*, 1–17.

Prusiner, S.B. (1998). Prions. Proc. Natl. Acad. Sci. U.S.A. *95*, 13363–13383.

Prusiner, S.B., Groth, D., Serban, A., Koehler, R., Foster, D., Torchia, M., Burton, D., Yang, S.L., and DeArmond, S.J. (1993). Ablation of the prion protein (PrP) gene in mice prevents scrapie and facilitates production of anti-PrP antibodies. Proc. Natl. Acad. Sci. U.S.A. *90*, 10608–10612.

Rice, J., Ottensmeier, C.H., and Stevenson, F.K. (2008). DNA vaccines: precision tools for activating effective immunity against cancer. Nat. Rev. Cancer *8*, 108–120.

Sakaguchi, S., Katamine, S., Shigematsu, K., Nakatani, A., Moriuchi, R., Nishida, N., Kurokawa, K., Nakaoke, R., Sato, H., Jishage, K., *et al.* (1995). Accumulation of proteinase K-resistant prion protein (PrP) is restricted by the expression level of normal PrP in mice inoculated with a mouse-adapted strain of the Creutzfeldt–Jakob disease agent. J. Virol. *69*, 7586–7592.

Sakaguchi, S., Ishibashi, D., and Matsuda, H. (2009). Antibody-based immunotherapeutic attempts in experimental animal models of prion diseases. Expert Opin. Ther. Pat. *19*, 907–917.

Sawada, M., Imai, F., Suzuki, H., Hayakawa, M., Kanno, T., and Nagatsu, T. (1998). Brain-specific gene expression by immortalized microglial cell-mediated gene transfer in the mammalian brain. FEBS Lett. *433*, 37–40.

Sethi, S., Lipford, G., Wagner, H., and Kretzschmar, H. (2002). Postexposure prophylaxis against prion disease with a stimulator of innate immunity. Lancet *360*, 229–230.

Sigurdsson, E.M., Brown, D.R., Daniels, M., Kascsak, R.J., Kascsak, R., Carp, R., Meeker, H.C., Frangione, B., and Wisniewski, T. (2002). Immunization delays the onset of prion disease in mice. Am. J. Pathol. *161*, 13–17.

Simard, A.R., and Rivest, S. (2004). Bone marrow stem cells have the ability to populate the entire central nervous system into fully differentiated parenchymal microglia. FASEB J. *18*, 998–1000.

Solforosi, L., Criado, J.R., McGavern, D.B., Wirz, S., Sanchez-Alavez, M., Sugama, S., DeGiorgio, L.A., Volpe, B.T., Wiseman, E., Abalos, G., *et al.* (2004). Cross-linking cellular prion protein triggers neuronal apoptosis *in vivo*. Science *303*, 1514–1516.

Song, C.H., Furuoka, H., Kim, C.L., Ogino, M., Suzuki, A., Hasebe, R., and Horiuchi, M. (2008). Effect of intraventricular infusion of anti-prion protein monoclonal antibodies on disease progression in prion-infected mice. J. Gen. Virol. *89*, 1533–1544.

Song, C.H., Honmou, O., Ohsawa, N., Nakamura, K., Hamada, H., Furuoka, H., Hasebe, R., and Horiuchi, M. (2009). Effect of transplantation of bone marrow-derived mesenchymal stem cells on mice infected with prions. J. Virol. *83*, 5918–5927.

Srivastava, P. (2002). Roles of heat-shock proteins in innate and adaptive immunity. Nat. Rev. Immunol. *2*, 185–194.

Thadani, V., Penar, P.L., Partington, J., Kalb, R., Janssen, R., Schonberger, L.B., Rabkin, C.S., and Prichard, J.W. (1988). Creutzfeldt–Jakob disease probably acquired from a cadaveric dura mater graft. Case report. J. Neurosurg. *69*, 766–769.

Wadsworth, J.D., Asante, E.A., Desbruslais, M., Linehan, J.M., Joiner, S., Gowland, I., Welch, J., Stone, L., Lloyd, S.E., Hill, A.F., *et al.* (2004). Human prion protein with valine 129 prevents expression of variant CJD phenotype. Science *306*, 1793–1796.

Wang, F., Wang, X., Yuan, C.G., and Ma, J. (2010). Generating a prion with bacterially expressed recombinant prion protein. Science *327*, 1132–1135.

Wells, G.A., Scott, A.C., Johnson, C.T., Gunning, R.F., Hancock, R.D., Jeffrey, M., Dawson, M., and Bradley, R. (1987). A novel progressive spongiform encephalopathy in cattle. Vet. Rec. *121*, 419–420.

Wernig, M., Zhao, J.P., Pruszak, J., Hedlund, E., Fu, D., Soldner, F., Broccoli, V., Constantine-Paton, M., Isacson, O., and Jaenisch, R. (2008). Neurons derived from reprogrammed fibroblasts functionally integrate into the fetal brain and improve symptoms of rats with Parkinson's disease. Proc. Natl. Acad. Sci. U.S.A. *105*, 5856–5861.

White, A.R., Enever, P., Tayebi, M., Mushens, R., Linehan, J., Brandner, S., Anstee, D., Collinge, J., and Hawke, S. (2003). Monoclonal antibodies inhibit prion replication and delay the development of prion disease. Nature *422*, 80–83.

Williams, E.S. (2005). Chronic wasting disease. Vet. Pathol. *42*, 530–549.

Wroe, S.J., Pal, S., Siddique, D., Hyare, H., Macfarlane, R., Joiner, S., Linehan, J.M., Brandner, S., Wadsworth, J.D., Hewitt, P., *et al.* (2006). Clinical presentation and pre-mortem diagnosis of variant Creutzfeldt–Jakob disease associated with blood transfusion: a case report. Lancet *368*, 2061–2067.

Wuertzer, C.A., Sullivan, M.A., Qiu, X., and Federoff, H.J. (2008). CNS delivery of vectored prion-specific single-chain antibodies delays disease onset. Mol. Ther. *16*, 481–486.

Yamanaka, H., Ishibashi, D., Yamaguchi, N., Yoshikawa, D., Nakamura, R., Okimura, N., Arakawa, T., Tsuji, T., Katamine, S., and Sakaguchi, S. (2006). Enhanced mucosal immunogenicity of prion protein following fusion with B subunit of *Escherichia coli* heat-labile enterotoxin. Vaccine *24*, 2815–2823.

Zugel, U., and Kaufmann, S.H. (1999). Role of heat shock proteins in protection from and pathogenesis of infectious diseases. Clin. Microbiol. Rev. *12*, 19–39.

Bovine Spongiform Encephalopathy and Scrapie

9

Takashi Yokoyama

Abstract

Bovine spongiform encephalopathy (BSE) emerged more than two decades ago, and to date over 190,000 head of cattle have been diagnosed. An effective feed ban programme has diminished the outbreak worldwide; however, different phenotypes of BSE (atypical BSEs) have emerged. Additionally, a different phenotype of scrapie (atypical scrapie) was reported in 1998. The prevalence of these diseases is low, suggesting the possibility that they are spontaneous forms of prion diseases. The decline in BSE outbreaks has led to discussion for the withdrawal of some control programmes. However, despite their unknown origin, the transmissibility of these atypical animal prion diseases has been demonstrated. Pre-emptive measures are necessary in order to prevent future outbreaks of atypical prion diseases in animals.

Bovine spongiform encephalopathy

Bovine spongiform encephalopathy (BSE) is categorized as transmissible spongiform encephalopathies (TSEs) or prion diseases. BSE was first reported in 1986 in the UK (Wells *et al.*, 1987), and over 190,000 cases have been reported since then. It is a fatal disease of cattle caused by prions, which mainly, if not entirely, consist of abnormal isoforms of prion protein (PrP^{Sc}) (Prusiner, 1991). PrP^{Sc} is generated by post-translational modification of normal cellular prion protein (PrP^{C}), and demonstrates partial resistance to proteinase digestion. This physicochemical difference is used in BSE diagnosis to discriminate PrP^{Sc} from PrP^{C}. Meat and bone meal (MBM), a feed supplement produced from animal carcasses and by-products, was used as a low-cost ingredient of animal feed. Though the origin of the BSE prion remains unknown, BSE prion-contaminated MBM eventually led to the circulation of the prions and caused the epidemic (Wilesmith *et al.*, 1988). Initial cases of BSE in some countries were considered the result of imported MBM and cattle from BSE-positive countries. Subsequently, domestically produced MBM derived from BSE-affected indigenous cattle amplified the infection. Thus, BSE is considered as a man-made disease.

Inter-species transmission of BSE

A wide range of host species is affected by BSE prions. In nature, BSE has been transmitted to cattle and several zoo ruminants (Kirkwood *et al.*, 1994). This disease was also transmitted to wild and domestic cats, causing feline spongiform encephalopathy (Hewicker-Trautwein and Bradley, 2006). Experimentally, BSE has been transmitted to mice (Fraser *et al.*, 1992), sheep, goats (Foster *et al.*, 1993), minks (Robinson *et al.*, 1994), marmosets (Baker *et al.*, 1998), macaques (Lasmezas *et al.*, 1996), and lemurs (Bons *et al.*, 1999). Some animal species have been observed to be resistant to BSE infection. While cats have been reported to develop feline spongiform encephalopathy, dogs do not appear to be affected by BSE, even when they are fed similar pet food (Kirkwood and Cunningham, 1994). Moreover, BSE prions are not transmitted to hamsters (Yokoyama *et al.*, 2009). Although the precise mechanisms of this species barrier phenomenon have not been fully elucidated, its biophysical basis is probably in the structural differences between host PrP^{C} and the PrP^{Sc} in the inoculum (Prusiner, 1991).

Until 1996, zoonotic potential of animal prion disease was not established. BSE was transmitted to humans and causing variant Creutzfeldt–Jakob disease (vCJD), 10 years after its first occurrence (Collinge *et al.*, 1996). The occurrence of vCJD in humans raised food safety concerns and categorized BSE as a zoonosis from a novel cattle disease. Human-to-human transmission of vCJD via transfusion has also been reported (Llewelyn *et al.*, 2004). Secondary transmission of vCJD by prion-contaminated medical instruments has also been suspected. The occurrence of vCJD forced the introduction of additional regulation in related medical fields (transfusion, transplantation, etc.). It also emphasized the importance of eradicating BSE in cattle.

Epidemiology

BSE cases have occurred worldwide. It appeared in the mid-1980s in the UK, and peaked in 1992. A ban on the use of MBM in ruminant feed resulted in a progressive decline of the epidemic. Because of the long incubation period of BSE, the effectiveness of the feed ban was demonstrated only 4–5 years after its introduction. However, as MBM-related feed-borne sources were the only substantiated route of infection, the feed ban on ruminant MBM to ruminants failed to eliminate BSE. Cross-contamination of cattle feed with pig and poultry feed, or cross-exposure of cattle feed to pig and/or poultry feed on farms have been suspected. It may have caused the relatively large numbers of BSE cases borne after the feed ban (BAB). The feeding ban of mammalian processed animal protein to cattle, sheep, and goats was introduced in 1994 in the EU. In 2001, the ban was expanded to encompass the feeding of all processed animal proteins to all farmed animals.

As MBM was exported from the UK, BSE was also spread to the EU countries, and the export of contaminated MBM from the EU then led to the spread of BSE to the non-EU countries (Fig. 9.1). Fig. 9.1 illustrates the spread of BSE from the UK to the EU, and then to the non-EU countries. Subsequently, the occurrence of BSE declined after the feed ban was introduced. Currently, the continuous decline of the BSE epidemic has been observed, indicating the effectiveness of the feed ban.

In Japan, the first case of BSE was detected in 2001, and to date 36 cases have been identified. Various control measures have since been taken to prevent human and animal exposure to BSE prions, including prohibition of the use of MBM in feed, removal of specified risk materials (SRMs: see below), testing of all slaughtered cattle at abattoirs, and fallen stock surveillance of all dead cattle aged over 24 months. Our experience showed that the detection of BSE-affected cattle numbers depended on the surveillance programme. The first BSE-affected cattle group was born in 1995 and 1996, and some cattle may have been processed into MBM before the occurrence of BSE in 2001, and caused the next BSE-affected generation born between 1999 and 2002 (Fig. 9.2).

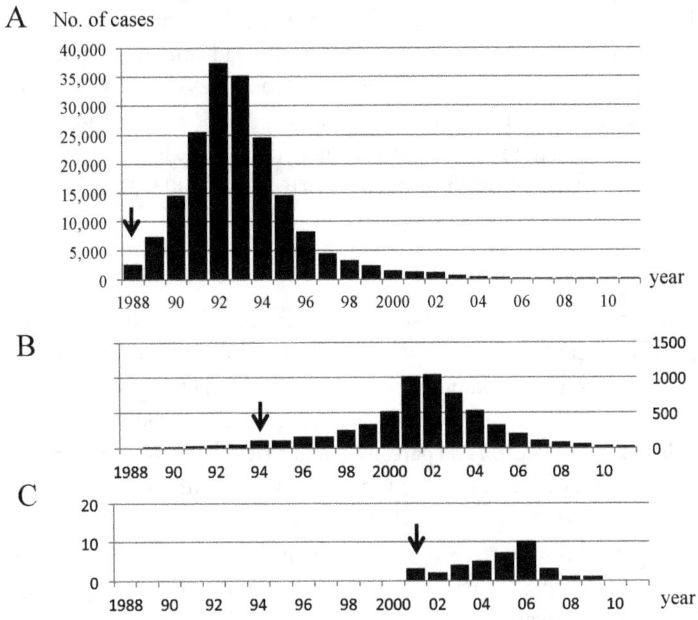

Figure 9.1 BSE statistics. BSE cases in the UK (A), the EU (excluding the UK) (B), and Japan (C). Arrows indicate when the feed ban programme was introduced.

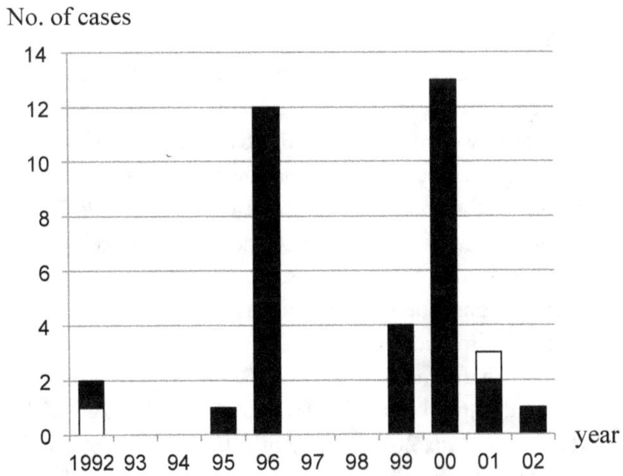

Figure 9.2 Birth year of BSE cattle in Japan. The birth year of each BSE case has been plotted. Black, C-BSE; white, L-BSE.

BSE prion strains

The uniform pathology among BSE-affected cattle and the limited results obtained from BSE transmission experiments in mice led to the consideration that BSE was caused by a single prion strain. Both BSE and its human counterpart, vCJD, have been associated with a single major prion strain. This strain is characterized by a unique and remarkably stable biochemical profile of PrP^{Sc} (dominant in diglycosylated PrP^{Sc} and lower molecular weight of unglycosylated PrP^{Sc}) isolated from the brains of affected animals or humans (Collinge *et al.*, 1996). The primary passage of BSE prions to different lines of inbred mice resulted in the isolation of two different prion strains with distinct PrP^{Sc} types, incubation periods, and different PrP^{Sc} deposition in the mice brains (Lloyd *et al.*, 2004). Furthermore, different BSE phenotypes, termed 'atypical BSE,' have been reported since 2003. These cases demonstrate the existence of different prion strains in BSE.

Clinical signs

BSE has a slowly progressive and invariably fatal course. It is characterized by a long incubation period in affected cattle, and the absence of a detectable immunological response in the host. BSE is characterized by (1) apprehension, behavioural changes, fear, increased startle response, or depression; (2) hyperreactivity or hyperreflexia to touch, sound and light; (3) gait ataxia, including hypermetria and paresis, resulting in falling; (4) adventitial movements such as muscle fasciculations, tremor, and myoclonus; (5) autonomic dysfunction including reduced rumination, bradycardia, and cardiac arrhythmia; and (6) loss of body weight, deterioration in general condition, and reduction in milk yield (Wells *et al.*, 1987; Wilesmith *et al.*, 1988; Braun *et al.*, 1998). These features can also arise in other central nervous system (CNS) disorders, and clinical features are not evident until the terminal stage of the illness.

Nervous ketosis, hypomagnesaemia, encephalic listeriosis and other encephalitides have been listed as differential diagnoses. It is difficult to diagnose BSE from clinical signs alone and laboratory tests are required for this, particularly to detect the preclinical stage.

Specified risk material

At the terminal stages of the disease, BSE prions are detected in the brain, spinal cord, retina, ileum, adrenal gland, trigeminal ganglion, thoracic ganglion, and peripheral nerves. At the preclinical stages, BSE prions were detected from the tonsils, bone marrow, and distal ileum of experimentally challenged cattle. As PrP^{Sc} mainly accumulates in the CNS during the middle to later stages of disease progression, none of the available BSE tests can detect affected cattle in the early stage of the disease. To address this problem, tissue that could harbour BSE prions are defined as SRM and removed from BSE-negative cattle to prevent consumption by humans. Consequent to the occurrence of vCJD, a risk-based approach has been adopted. Recently, limited amounts of PrP^{Sc} and/or infectivity were detected in parts of the intestine other than the distal ileum in BSE-infected cattle under experimental inoculation (jejunum) and natural cases (distal jejunum and colon) (Okada *et al.*, 2010; Hoffmann *et al.*, 2011; Stack *et al.*, 2011b). The intestines spanning the duodenum to the rectum of cattle of all ages are currently included in the EU list of SRM (EFSA, 2011).

Diagnosis

PrP^{Sc} accumulates in animals affected with TSE. The detection of PrP^{Sc} and the differentiation of this isoform from PrP^{C} are crucial for the diagnosis of TSE. The conversion of PrP^{C} to PrP^{Sc} is the central event in prion propagation; however, the mechanism of this conformational change remains obscure.

The diagnosis of BSE has been performed using histopathology (Wells *et al.*, 1987), immunohistochemistry (IHC) (Haritani *et al.*, 1994), and biochemical analyses such as Western blot (WB) (Hope *et al.*, 1988; Grassi *et al.*, 2001) and enzyme-linked immunosorbent assay

(ELISA)-based tests (Schaller *et al.*, 1999; Deslys *et al.*, 2001). Other than histological examination, these test kits aim to measure PrP^{Sc} as proteinase K (PK)-resistant PrP. PrP^{Sc} is a conformational isomer of the host protein, thus prion infection does not induce a significant immune response in the host. No serological tests are available for diagnosis.

The methods of BSE diagnosis are all post-mortem procedures for detecting PrP^{Sc}, and are applicable only in brain tissue (obex region). Several commercial kits are available for BSE diagnosis and have been introduced for both active (healthy slaughtered cattle) and passive surveillance (dead-on-farm cattle), contributing to BSE control. Large-scale biochemical analysis of PrP^{Sc} in BSE has led to the detection of atypical BSE cases (see below).

Recently, protein misfolding cyclic amplification (PMCA), which is a highly sensitive technique for detecting minute amounts of PrP^{Sc}, enabled the detection of PrP^{Sc} in BSE-affected cattle tissue (Murayama *et al.*, 2010). The distribution of PrP^{Sc} was not restricted to nervous tissue, and could spread to peripheral tissue (e.g. saliva, submandibular gland, and sublingual gland) in the terminal diseased cattle (Murayama *et al.*, 2010). This technique is expected to evaluate the safety of livestock products, raw feed materials and bovine-source materials.

Atypical BSE

Based on the analyses of its biological and biochemical characteristics, BSE (hereafter referred as classical BSE: C-BSE) was thought to be caused by a single prion strain (Bruce *et al.*, 1994). Since 2003, however, several atypical neuropathological and molecular phenotypes of BSE (atypical BSE) have been detected in Japan, several European countries, and North America (Yamakawa *et al.*, 2003; Biacabe *et al.*, 2004; Casalone *et al.*, 2004; Hagiwara *et al.*, 2007; Jacobs *et al.*, 2007; Richt *et al.*, 2007; Terry *et al.*, 2007; Tester *et al.*, 2009; Dudas *et al.*, 2010), totalling more than 60 cases worldwide (Balkema-Buschmann *et al.*, 2011). The exact N-terminal sequence of the PrP^{Sc} PK-digestion site depends on the particular prion strain. The PK cleavage sites of C-BSE prions differ from those of scrapie prions (Fig. 9.3) (Hayashi *et al.*, 2005). Based on the molecular size, glycoform ratio and the antibody-binding characteristics of PrP^{Sc}, atypical BSE has different PrP^{Sc} characteristics and is currently classified into two groups, designated L-BSE and H-BSE (Jacobs *et al.*, 2007). L-BSE, also known as bovine amyloidotic spongiform encephalopathy (BASE), which was identified in Italy (Casalone *et al.*, 2004). A spontaneous form of cattle prion disease (Jacobs *et al.*, 2007) or a genetic origin associated with an E211K mutation in the H-BSE case from the USA (Nicholson *et al.*, 2008) have been cited in relation to the origin of atypical BSEs. Recently, a novel PrP^{Sc} phenotype distinct from those of C-, L-, and H-BSE was reported in aged cows in Switzerland (Seuberlich *et al.*, 2012). This case was detected with active surveillance, and IHC was inconclusive because of severe tissue autolysis. Further transmission studies are expected conclude whether it is another atypical BSE.

PrP^{Sc} profile of atypical BSEs

H-BSE is characterized by a significantly higher unglycosylated PrP^{Sc} molecular mass; however, it possesses a conventional glycopattern (diglycosylated PrP^{Sc}-dominant). Further truncated C-terminal PrP fragments are detected in PrP^{Sc} of H-BSE. It means that different PrP^{Sc} conformers coexist in H-BSE-affected animal. In contrast, the unglycosylated PrP^{Sc} of L-BSE has a slightly lower molecular mass, with a distinctly different glycoprofile. Thus,

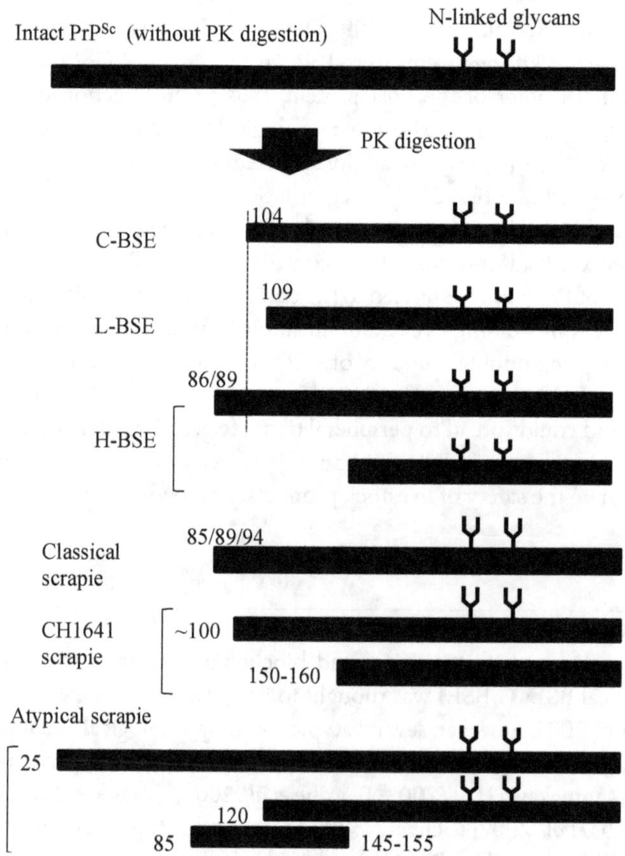

Figure 9.3 PrP^{Sc} fragments after PK digestion. Intact PrP^{Sc} is shown at the top. After PK digestion, truncated PrP^{Sc} bands were observed. C-, L-, and H-BSE have different N-terminal cleavage sites. H-BSE has an additional truncated PrP^{Sc} band. A similar band was detected in the CH1641 scrapie. PrP^{Sc} from atypical scrapie is clearly different from that in other TSEs.

distinct PrP^{Sc} characteristics are observed in atypical BSEs, and this may link to the biological differences of BSE prions. Immunoreactivity against the N-terminal end of PrP^{Sc} could discriminate L-BSE from C-BSE and H-BSE (Fig. 9.3) (Stack *et al.*, 2011a).

Transmissibility of atypical BSEs

Cattle affected with atypical BSEs demonstrate disease phenotypes distinct from that of C-BSE. Cattle inoculated with L-BSE prions exhibited progressive muscle atrophy and behavioural changes characterized by weakness, severe lethargy, and ataxia (Lombardi *et al.*, 2008). H-BSE was also successfully transmitted to calves with incubation periods between 500 and 600 days. The infected animals showed loss of body condition, ataxia of the forelimbs and hindlimbs and myoclonus and were unable to rise. However, animals did not show any change in temperament, such as nervousness or aggression (Okada *et al.*, 2011a).

The incubation periods of atypical BSE-challenged cattle varied with each experiment. The incubation period of BASE-inoculated Friesian cattle was shorter than that of Alpine brown cattle (Lombardi *et al.*, 2008). This breed-associated effect suggests the existence of a disease-modifier other than PrP^C (Lombardi *et al.*, 2008). The incubation periods of L-BSE in the Holstein breed were shorter than that of C-BSE (Table 9.1) (Fukuda *et al.*, 2009).

The clinical signs of atypical BSE-challenged cattle (Holstein–Friesian) have been summarized in Table 9.1 (Fukuda *et al.*, 2009; Okada *et al.*, 2011a, 2012).

Cattle experimentally infected with H-BSE harbour PrP^{Sc} plaques and glial- and stellate-type PrP^{Sc} accumulation (Okada *et al.*, 2011a). L-BSE also demonstrates a different distribution and topology of PrP^{Sc} accumulation, featuring PrP-immunopositive amyloid plaques (Casalone *et al.*, 2004; Fukuda *et al.*, 2009). The presence of PrP^{Sc} plaques in the forebrain, but not the brainstem, is a neuropathological feature in atypical BSE-affected cattle (Casalone *et al.*, 2004; Okada *et al.*, 2012). The PrP^{Sc} accumulation patterns or topology of atypical BSEs also differ from that of C-BSE. However, a pinpoint IHC analysis of the medulla oblongata, which is used for routine BSE diagnosis, could not discriminate atypical BSE from C-BSE (Konold *et al.*, 2012).

The results of cattle orally challenged with atypical BSEs are expected to evaluate the SRM definition of atypical BSE in cattle. PrP^{Sc} accumulation in cattle intracerebrally challenged with atypical BSEs resembled that of C-BSE cattle. PrP^{Sc} has been detected from CNS tissue, peripheral tissue, and the adrenal gland, but not from lymphoid tissue (Iwamaru *et al.*, 2010). These data show that the current SRM definition of cattle is also applicable to L-BSE-affected cattle.

The transmissibility of atypical BSEs was also confirmed with transgenic (Tg) mice models. L-BSE-inoculated bovine PrP Tg mice had significantly shorter incubation periods than C-BSE–inoculated mice (Buschmann *et al.*, 2006; Masujin *et al.*, 2008) (Table 9.1). H-BSE-inoculated TgBoPrP mice had longer incubation periods than the C-BSE-inoculated mice. The incubation period of H-BSE-challenged cattle was shorter than that of C-BSE-challenged cattle (Table 9.1). The discrepancy in this result suggests that the incubation periods and/or the susceptibility may influence the unidentified host factors, other than the cattle PrP.

Table 9.1 Summary of C-, L-, and H-BSE transmission to cattle and bovine PrP Tg mice

	BSE prions	Diseased/ inoculated	Incubation period (days)	Clinical signs/ PrP^{Sc} phenotype	Reference
Cattle (Holstein–Friesian)	C-BSE	3/3	675±57	Ataxic gait, uncoordinated hind limbs, hyper-sensitive	Fukuda *et al.* (2009)
	L-BSE	3/3	486±11	Ataxic gait, uncoordinated hind limbs, inactive, little aggression	Fukuda *et al.* (2009)
	H-BSE	3/3	560±47	Ataxia of the forelimbs and hind limbs, myoclonus	Okada *et al.* (2011a)
TgBoPrP	C-BSE	11/11	223.5±13.5	C-BSE phenotype	Masujin *et al.* (2008)
	L-BSE	10/10	197.7±3.4	L-BSE phenotype	Masujin *et al.* (2008)
	H-BSE	10/10	315.8±11.6	H-BSE phenotype	Okada *et al.* (2011c)

TgBoPrP, bovine PrP Tg mice.

The distinct traits of atypical BSE are maintained in intracerebral inoculation to cattle and bovine PrP Tg mice (Tgbov XV, TgBoPrP) (Scott *et al.*, 1999; Buschmann *et al.*, 2006; Masujin *et al.*, 2008; Okada *et al.*, 2011b), differing from those of C-BSE. This indicates that atypical BSEs are caused by prion strains different from those that cause C-BSE. However, small numbers of different H-BSE-inoculated bovinized Tg mice lines (Tg110) exhibited C-BSE–like phenotypes (Torres *et al.*, 2011). The majority of wild-type mice (C57BL/6), into which H-BSE was transmitted, also exhibited the H-BSE phenotype, and small numbers of mice exhibited the C-BSE-like phenotype (Baron *et al.*, 2011). Similarly, L-BSE prions acquired C-BSE–like characteristics through interspecies transmission in wild-type mice (Capobianco *et al.*, 2007) and ovine PrP Tg mice (Beringue *et al.*, 2007). These results gave rise to the hypothesis that C-BSE could have originated from atypical BSEs during inter- and/or intraspecies transmission.

Risk for humans

The transmission rate and incubation period in human PrP Tg mice inoculated intracerebrally with L-BSE are higher and shorter than those of C-BSE-challenged mice, respectively (Beringue *et al.*, 2008a; Kong *et al.*, 2008). Furthermore, macaques challenged intracerebrally with L-BSE exhibited the clinical signs of disease after 26 months: this period was shorter than that of C-BSE-affected macaques; however, it was similar to that of vCJD-challenged animals (Comoy *et al.*, 2008; Ono *et al.*, 2011). L-BSE could be transmitted to mouse lemurs by oral challenge (Mestre-Frances *et al.*, 2012). Although little is known about their epidemiology, pathobiology, and zoonotic potential, L-BSE prions appear more virulent than C-BSE prions in humans.

In contrast, attempts to transmit H-BSE to human PrP Tg mice have failed (Beringue *et al.*, 2008a). In addition, successful transmission of H-BSE prions to non-human primates has not been reported.

Risk for animals

It has been reported that the host range of the L-BSE prions was different from that of the C-BSE prions (Capobianco *et al.*, 2007; Masujin *et al.*, 2008; Nicot and Baron, 2011; Shu *et al.*, 2011). However, the susceptible host range of atypical BSEs is unclear. Transmissible mink encephalopathy (TME) is a prion disease in mink in North America and Europe. Although feed contaminated with scrapie sheep tissue has been proposed as the origin of TME, the origin of the disease remains elusive. Epidemiological analysis of cases in Stetsonville, Wisconsin, and other US cases of TME have proposed a possible cattle origin for TME. Mink are not fed sheep products; however, they are fed deer or diseased (downer) cattle carcasses (Marsh *et al.*, 1991; Marsh and Bessen, 1993). It has also been proposed that the Stetsonville TME prion is distinct from that of C-BSE, but has phenotypic similarities to L-BSE in ovinized Tg mice (Baron *et al.*, 2007a). A hypothesis proposes that L-BSE is a candidate for a bovine source of TME (Baron *et al.*, 2007a). Clarification of the susceptible host range of atypical BSEs is important for risk analysis.

Conclusion

While C-BSE is thought to be the result of feeding with prion-contaminated MBM, the origin of atypical BSEs remains unknown. The incidence of atypical BSE is very low (sporadic occurrence), and cases are mainly detected in aged cattle (Biacabe *et al.*, 2008). Several hypotheses regarding its origin have been proposed that suggest a major change in the BSE agent or another prion disease such as scrapie, which affects sheep and goats. Some suggest that it is a previously unrecognized prion disease in cattle or that it is a spontaneous form of prion disease.

Atypical BSEs have proven to be transmissible despite their obscure origin. The occurrence of C-BSE has declined because of the effectiveness of the feed ban programme. However, in order to prevent future outbreaks of atypical BSEs, the present BSE control programmes should not be withdrawn without careful consideration. Furthermore, studies on the risk analysis of atypical BSEs are required.

Scrapie

Scrapie in sheep and goats is the longest known and most widespread TSE. Its worldwide occurrence in indigenous sheep has been recognized for more than 250 years. Although Australia and New Zealand appear to be free from scrapie, the disease status of many countries is unknown because of the lack of sufficient surveillance. The discovery and endemic occurrence of BSE had drawn attention to animal prion diseases. Subsequently, scrapie became a notified disease in the EU in 1993.

Scrapie occurs mostly in sheep aged 2–5 years, and there is no upper age limit. The disease affects most sheep breeds. The susceptibility of sheep to scrapie is strongly associated with alleles of the PrP gene and with polymorphisms at codons 136, 154, and 171 (Dawson *et al.*, 1998). For example, sheep scrapie is frequently found to be correlated with homozygosity QQ at codon 171. However, a limited number of reports on scrapie in genetically resistant sheep suggest that the resistance is relative rather than absolute (Ikeda *et al.*, 1995; Groschup *et al.*, 2007). Despite the uncertainty of whether the resistance is absolute, or simply prevents the onset of clinical diseases, large-scale genotyping and breeding programmes of scrapie-resistant sheep flocks were conducted in many EU countries (Dawson *et al.*, 1998; Arnold *et al.*, 2002). However, atypical scrapie (see below) in ARR/ARR sheep that were selected to generate scrapie-resistant flocks is a caution against the aim of the genetic selection programme.

Clinical signs

Prion pathology varies with the host species. PrP amino acid polymorphisms are attributed to the susceptibility to scrapie and to the incubation periods (Westaway *et al.*, 1987, 1994; Lee *et al.*, 2001). Scrapie is recognized based on behavioural changes and progressive nervous alterations: incoordination, abnormal gait, pruritus with rubbing against objects, and loss of wool. The clinical phase lasts several weeks to several months. However, some animals die suddenly without apparent clinical signs.

PrP^{Sc} distribution

The disease phenotype of scrapie sheep is different from BSE-affected cattle. PrP^{Sc} is detected from the CNS, tonsils, spleen, lymph nodes, nictitating membrane, muscles, placenta, distal ileum, and proximal colon in scrapie-affected sheep. PrP^{Sc} detection from CNS or lymphoid tissues by ELISA, WB, and/or IHC has been used for diagnosis (Ikegami *et al.*, 1991; Race *et al.*, 1992; van Keulen *et al.*, 1996). Detection of PrP^{Sc} from tonsils (Schreuder *et al.*, 1996; Shimada *et al.*, 2005), the nictitating membrane (O'Rourke *et al.*, 2000), and rectoanal mucosa-associated lymphoid tissue (Dennis *et al.*, 2009) is used as a diagnostic test in live animals.

It is believed that most animals are infected at birth or shortly thereafter. It has been reported that placenta from infected ewes could transmit disease by the oral route. In addition, an environment contaminated with scrapie prions (such as from placenta) may infect sheep. In addition to the oral route of exposure, the conjunctiva and broken skin are also suspected routes of transmission (Hornlimann *et al.*, 2006).

Scrapie prion strains

Scrapie prions are classified into many different strains based on the incubation period, lesion profile, and PrP^{Sc} distribution in inbred mice. Conformational differences in the PrP^{Sc} structure may contribute to strain variations (Bessen and Marsh, 1994; Telling *et al.*, 1996; Legname *et al.*, 2005). The heterogeneity of natural sheep scrapie is reflected in the results of PrP^{Sc} molecular profiling (Owen *et al.*, 2007). Scrapie was believed to be the origin of BSE; however, a BSE-like scrapie strain has not been detected yet. Multiple prion strains have been isolated from some cases of natural sheep scrapie (Kimberlin and Walker, 1978; Masujin *et al.*, 2009), and information regarding the pathogenesis of scrapie prion strains in their original host, sheep, is limited. In field sheep scrapie, the high diversity of strains present appears to favour the multiple conformer concept (Bruce and Dickinson, 1987; Bruce, 1993; Bruce *et al.*, 2002).

One of the more unusual scrapie isolates, CH1641, and the PrP^{Sc} that causes BSE have similar molecular weights (Yokoyama *et al.*, 2010), being lower than the molecular weight of the PrP^{Sc} of typical scrapie (Baron *et al.*, 2004; Lezmi *et al.*, 2004; Stack *et al.*, 2006; Baron and Biacabe, 2007). Furthermore, CH1641 prions contain a 14-kDa protein fragment produced by the C-terminal cleavage of PrP^{Sc} (Baron *et al.*, 2008). A transmission study in ovine Tg mice revealed differences between CH1641 scrapie and BSE (Baron *et al.*, 2008). Natural cases of CH1641-like scrapie, a rare disease, also have been reported (Lezmi *et al.*, 2004; Stack *et al.*, 2006).

In experimental transmission of typical scrapie, two different scrapie phenotypes were observed in different sheep breeds, even though they had identical PrP genotypes (Yokoyama *et al.*, 2010). One was similar to CH1641-like scrapie, and the other was similar to the PrP^{Sc} in the original inocula. In the case of CH1641 prions, PrP^{Sc} mainly accumulate in the neurons of the affected sheep, and very little extracellular PrP^{Sc} accumulation is observed in IHC (Jeffrey *et al.*, 2006). It has been reported that the scrapie strain in sheep might influence the resulting PrP phenotype, and PrP^{Sc} accumulation in lymphoid tissue is unrelated to the route of infection (Siso *et al.*, 2010). This indicates that prion propagation is influenced by the host species (Yokoyama *et al.*, 2010).

In interspecies transmission of sheep scrapie to mice, the proportional change of the PrP^{Sc} conformer was consistent with the stabilization of the incubation period and the change in the biochemical characteristics of PrP^{Sc} (Ushiki-Kaku *et al.*, 2010). A molecular-level model of PrP^{Sc} conformation in interspecies transmission has been proposed (Collinge and Clarke, 2007; Beringue *et al.*, 2008b). In this model, the PrP^{Sc} of one strain is believed to represent a cluster of several conformers. Among these conformers, one or more PrP^{Sc} conformers that were most readily adaptable in the host were selected; these conformers then propagated dominantly to become host-adapted PrP^{Sc}. Another possibility is that mutation of uniform-conformation PrP^{Sc} causes the emergence of new, host-adapted PrP^{Sc}.

Prion strain diversity and TSE pathogenesis in sheep should be studied, and the findings obtained may help reveal the origin of BSE.

Atypical scrapie

In 1998, a different phenotype of sheep prion disease (termed atypical scrapie; also known as Nor98) was diagnosed in Norway (Benestad *et al.*, 2003). Following the implementation of a surveillance programme for TSE in small ruminants, increasing numbers of atypical scrapie cases have been reported in most EU countries. Active surveillance has detected many atypical scrapie cases in apparently healthy sheep or fallen stock animals, and not in clinical suspect cases. Furthermore, atypical scrapie has been detected in sheep in the Falkland Islands, USA, Canada (Mitchell *et al.*, 2010), UK (Griffiths *et al.*, 2010), and New Zealand (Kittelberger *et al.*, 2010). It is now considered a worldwide disease.

In most EU countries, the prevalence of atypical scrapie is low. The prevalence estimates of classical scrapie are more variable than that for atypical scrapie. The aetiology of atypical scrapie differs from that of classical scrapie. It has proposed that atypical scrapie may be sporadic, or if it is infectious, the rate of infection is very low in natural circumstances. It may also be influenced by genetic and metabolic factors (Fediaevsky *et al.*, 2009).

Disease phenotype

The mean age of affected sheep is 6.5 years. The ages of atypical scrapie cases appear to be higher than that of classical scrapie cases (Benestad *et al.*, 2003). Detecting only a single atypical scrapie case in the flock is common, and few secondary cases have been reported (Benestad *et al.*, 2008). Most of the affected sheep carry PrP alleles associated with resistance to classical scrapie. Some rapid diagnostic test results based on PK-resistant PrP^{Sc} detection have demonstrated discrepancies (Arsac *et al.*, 2007). PrP^{Sc} from atypical scrapie has been demonstrated to have lower protease resistance, which might explain the discrepancies between diagnostic tests (Baron *et al.*, 2007b).

Ataxia, anxiety, and loss of body condition have been reported as the silent clinical signs of atypical scrapie (Benestad *et al.*, 2008). Some cases do not exhibit any clinical signs otherwise. This might suggest that the clinical signs, if present, can differ or are less pronounced than that of classical scrapie (Benestad *et al.*, 2008).

The high susceptibility of atypical scrapie appears to be associated with the AHQ and AF141RQ alleles, whereas association with the VRQ allele is rare (Benestad *et al.*, 2003; Arsac *et al.*, 2007). The ARR/ARR genotype, which is known as a scrapie-resistant breed, is susceptible to atypical scrapie (Buschmann *et al.*, 2004; Arsac *et al.*, 2007).

PrP^{Sc} deposition

In sheep with atypical scrapie, fine punctate to coarse granular PrP^{Sc} deposition has been observed in grey and white matter. However, intraneuronal PrP^{Sc} accumulation, which is a typical characteristic of classical scrapie, has not been observed. PrP^{Sc} accumulation in the medulla oblongata is much less; however, there is greater PrP^{Sc} accumulation in the cerebellar and cerebral cortices of atypical scrapie sheep (Benestad *et al.*, 2008). This means that test results may depend on which part of the brain is examined.

Transmissibility of atypical scrapie

The transmissibility of atypical scrapie to ovine PrP Tg mice (Arsac *et al.*, 2009) and to sheep (Simmons *et al.*, 2011) has been confirmed. Furthermore, it is transmissible to porcine PrP Tg mice, acquiring new strain properties in the process (Espinosa *et al.*, 2009). Interestingly, porcine PrP Tg mice demonstrated resistance against classical scrapie infection. Furthermore, atypical scrapie is not transmissible to wild-type mice (Griffiths *et al.*, 2010). These results mean that the susceptible host range of atypical scrapie differs from that of classical scrapie.

Atypical scrapie has been experimentally transmitted to sheep orally. The disease phenotype is preserved upon sub-passage in sheep (Simmons *et al.*, 2007, 2010). PrP^{Sc} was detected in the brain, and while the peripheral tissues harboured infectivity, but were negative for PrP^{Sc} in the affected sheep (Simmons *et al.*, 2011).

PrP^{Sc} profile

The PrP^{Sc} band pattern of atypical scrapie differs from those of other prion diseases. The PrP^{Sc} produces multiple banding patterns, with molecular weights of approximately 7, 14 and 23 kDa (Gretzschel *et al.*, 2006; Arsac *et al.*, 2007; Baron *et al.*, 2007b; Benestad *et al.*, 2008). The 7-kDa fragment corresponds to PrP85–148, which results from proteolysis at both the N- and C-terminal ends of the PrP^{Sc}. Another 14-kDa fragment, PrP120–223, and its glyco-modified forms generate a laddered banding pattern (Fig. 9.3).

Conclusion

The co-existence of classical scrapie and atypical scrapie has been reported in Italian sheep (Mazza *et al.*, 2010). The PrP^{Sc} of classical scrapie accumulated in the brain stem, and that of atypical scrapie in the cerebral cortex and cerebellum (Benestad *et al.*, 2008). The areas targeted by the classical and atypical scrapie prions differed; hence, classical scrapie prions and atypical scrapie prions have different biological characteristics. In addition, it means that obex sampling may overlook the existence of atypical scrapie, diagnosing only classical scrapie.

The origin of atypical scrapie is unknown. A spontaneous, non-contagious origin such as sporadic CJD (sCJD) in humans cannot be excluded (Benestad *et al.*, 2008). A novel type of sporadic prion disease in humans, termed protease-sensitive prionopathy, has been reported (Gambetti *et al.*, 2008). Its PK-sensitive PrP profile is similar to that of atypical scrapie. Comparative study of these atypical prions may facilitate the definition of PrP^{Sc} and/or prions.

References

Arnold, M., Meek, C., Webb, C.R., and Hoinville, L.J. (2002). Assessing the efficacy of a ram-genotyping programme to reduce susceptibility to scrapie in Great Britain. Prev. Vet. Med. *56*, 227–249.

Arsac, J.N., Andreoletti, O., Bilheude, J.M., Lacroux, C., Benestad, S.L., and Baron, T. (2007). Similar biochemical signatures and prion protein genotypes in atypical scrapie and Nor98 cases, France and Norway. Emerg. Infect. Dis. *13*, 58–65.

Arsac, J.N., Betemps, D., Morignat, E., Feraudet, C., Bencsik, A., Aubert, D., Grassi, J., and Baron, T. (2009). Transmissibility of atypical scrapie in ovine transgenic mice: major effects of host prion protein expression and donor prion genotype. PLoS One *4*, e7300.

Baker, H.F., Ridley, R.M., Wells, G.A., and Ironside, J.W. (1998). Prion protein immunohistochemical staining in the brains of monkeys with transmissible spongiform encephalopathy. Neuropathol. Appl. Neurobiol. *24*, 476–486.

Balkema-Buschmann, A., Ziegler, U., McIntyre, L., Keller, M., Hoffmann, C., Rogers, R., Hills, B., and Groschup, M.H. (2011). Experimental challenge of cattle with German atypical bovine spongiform encephalopathy (BSE) isolates. J. Toxicol. Environ. Health A *74*, 103–109.

Baron, T., and Biacabe, A.G. (2007). Molecular behaviors of 'CH1641-like' sheep scrapie isolates in ovine transgenic mice (TgOvPrP4). J. Virol. *81*, 7230–7237.

Baron, T., Crozet, C., Biacabe, A.G., Philippe, S., Verchere, J., Bencsik, A., Madec, J.Y., Calavas, D., and Samarut, J. (2004). Molecular analysis of the protease-resistant prion protein in scrapie and bovine spongiform encephalopathy transmitted to ovine transgenic and wild-type mice. J. Virol. *78*, 6243–6251.

Baron, T., Bencsik, A., Biacabe, A.G., Morignat, E., and Bessen, R.A. (2007a). Phenotypic similarity of transmissible mink encephalopathy in cattle and L-type bovine spongiform encephalopathy in a mouse model. Emerg. Infect. Dis. *13*, 1887–1894.

Baron, T., Biacabe, A.G., Arsac, J.N., Benestad, S., and Groschup, M.H. (2007b). Atypical transmissible spongiform encephalopathies (TSEs) in ruminants. Vaccine *25*, 5625–5630.

Baron, T., Bencsik, A., Vulin, J., Biacabe, A.G., Morignat, E., Verchere, J., and Betemps, D. (2008). A C-terminal protease-resistant prion fragment distinguishes ovine 'CH1641-like' scrapie from bovine classical and L-Type BSE in ovine transgenic mice. PLoS Pathog. *4*, e1000137.

Baron, T., Vulin, J., Biacabe, A.G., Lakhdar, L., Verchere, J., Torres, J.M., and Bencsik, A. (2011). Emergence of classical BSE strain properties during serial passages of H-BSE in wild-type mice. PLoS One *6*, e15839.

Benestad, S.L., Sarradin, P., Thu, B., Schonheit, J., Tranulis, M.A., and Bratberg, B. (2003). Cases of scrapie with unusual features in Norway and designation of a new type, Nor98. Vet. Rec. *153*, 202–208.

Benestad, S.L., Arsac, J.N., Goldmann, W., and Noremark, M. (2008). Atypical/Nor98 scrapie: properties of the agent, genetics, and epidemiology. Vet. Res. *39*, 19.

Beringue, V., Andreoletti, O., Le Dur, A., Essalmani, R., Vilotte, J.L., Lacroux, C., Reine, F., Herzog, L., Biacabe, A.G., Baron, T., *et al.* (2007). A bovine prion acquires an epidemic bovine spongiform encephalopathy strain-like phenotype on interspecies transmission. J. Neurosci. *27*, 6965–6971.

Beringue, V., Herzog, L., Reine, F., Le Dur, A., Casalone, C., Vilotte, J.L., and Laude, H. (2008a). Transmission of atypical bovine prions to mice transgenic for human prion protein. Emerg. Infect. Dis. *14*, 1898–1901.

Beringue, V., Vilotte, J.L., and Laude, H. (2008b). Prion agent diversity and species barrier. Vet. Res. *39*, 47.

Bessen, R.A., and Marsh, R.F. (1994). Distinct PrP properties suggest the molecular basis of strain variation in transmissible mink encephalopathy. J. Virol. *68*, 7859–7868.

Biacabe, A.G., Laplanche, J.L., Ryder, S., and Baron, T. (2004). Distinct molecular phenotypes in bovine prion diseases. EMBO Rep. *5*, 110–115.

Biacabe, A.G., Morignat, E., Win, J., Calavas, D., and Baron, T.G.M. (2008). Atypical bovine spongiform encephalopathies, France, 2001–2007. Emerg. Infect. Dis. *14*, 298–300.

Bons, N., Mestre-Frances, N., Belli, P., Cathala, F., Gajdusek, D.C., and Brown, P. (1999). Natural and experimental oral infection of nonhuman primates by bovine spongiform encephalopathy agents. Proc. Natl. Acad. Sci. U.S.A. *96*, 4046–4051.

Braun, U., Schicker, E., and Hornlimann, B. (1998). Diagnostic reliability of clinical signs in cows with suspected bovine spongiform encephalopathy. Vet. Rec. *143*, 101–105.

Bruce, M.E. (1993). Scrapie strain variation and mutation. Brit. Med. Bull. *49*, 822–838.

Bruce, M.E., and Dickinson, A.G. (1987). Biological evidence that scrapie agent has an independent genome. J. Gen. Virol. *68*, 79–89.

Bruce, M., Chree, A., Mcconnell, I., Foster, J., Pearson, G., and Fraser, H. (1994). Transmission of bovine spongiform encephalopathy and scrapie to mice – Strain variation and the species barrier. Philos. T. Roy. Soc. B *343*, 405–411.

Bruce, M.E., Boyle, A., Cousens, S., McConnell, I., Foster, J., Goldmann, W., and Fraser, H. (2002). Strain characterization of natural sheep scrapie and comparison with BSE. J. Gen. Virol. *83*, 695–704.

Buschmann, A., Biacabe, A.G., Ziegler, U., Bencsik, A., Madec, J.Y., Erhardt, G., Luhken, G., Baron, T., and Groschup, M.H. (2004). Atypical scrapie cases in Germany and France are identified by discrepant reaction patterns in BSE rapid tests. J. Virol. Methods *117*, 27–36.

Buschmann, A., Gretzschel, A., Biacabe, A.G., Schiebel, K., Corona, C., Hoffmann, C., Eiden, M., Baron, T., Casalone, C., and Groschup, M.H. (2006). Atypical BSE in Germany-proof of transmissibility and biochemical characterization. Vet. Microbiol. *117*, 103–116.

Capobianco, R., Casalone, C., Suardi, S., Mangieri, M., Miccolo, C., Limido, L., Catania, M., Rossi, G., Di Fede, G., Giaccone, G., *et al.* (2007). Conversion of the BASE prion strain into the BSE strain: the origin of BSE? PLoS Pathog. 3, e31.

Casalone, C., Zanusso, G., Acutis, P., Ferrari, S., Capucci, L., Tagliavini, F., Monaco, S., and Caramelli, M. (2004). Identification of a second bovine amyloidotic spongiform encephalopathy: molecular similarities with sporadic Creutzfeldt–Jakob disease. Proc. Natl. Acad. Sci. U.S.A. *101*, 3065–3070.

Collinge, J., and Clarke, A.R. (2007). A general model of prion strains and their pathogenicity. Science *318*, 930–936.

Collinge, J., Sidle, K.C., Meads, J., Ironside, J., and Hill, A.F. (1996). Molecular analysis of prion strain variation and the aetiology of 'new variant' CJD. Nature *383*, 685–690.

Comoy, E.E., Casalone, C., Lescoutra-Etchegaray, N., Zanusso, G., Freire, S., Marce, D., Auvre, F., Ruchoux, M.M., Ferrari, S., Monaco, S., *et al.* (2008). Atypical BSE (BASE) transmitted from asymptomatic aging cattle to a primate. PLoS One *3*, e3017.

Dawson, M., Hoinville, L.J., Hosie, B.D., and Hunter, N. (1998). Guidance on the use of PrP genotyping as an aid to the control of clinical scrapie. Scrapie Information Group. Vet. Rec. *142*, 623–625.

Dennis, M.M., Thomsen, B.V., Marshall, K.L., Hall, S.M., Wagner, B.A., Salman, M.D., Norden, D.K., Gaiser, C., and Sutton, D.L. (2009). Evaluation of immunohistochemical detection of prion protein in rectoanal mucosa-associated lymphoid tissue for diagnosis of scrapie in sheep. Am. J. Vet. Res. *70*, 63–72.

Deslys, J.P., Comoy, E., Hawkins, S., Simon, S., Schimmel, H., Wells, G., Grassi, J., and Moynagh, J. (2001). Public health – screening slaughtered cattle for BSE. Nature *409*, 476–478.

Dudas, S., Yang, J., Graham, C., Czub, M., McAllister, T.A., Coulthart, M.B., and Czub, S. (2010). Molecular, biochemical and genetic characteristics of BSE in Canada. PLoS One *5*, e10638.

EFSA. (2011). Scientific opinion on a review of the BSE-related risk in bovine intestines. EFSA J. *9(3)*, 2104.

Espinosa, J.C., Herva, M.E., Andreoletti, O., Padilla, D., Lacroux, C., Cassard, H., Lantier, I., Castilla, J., and Torres, J.M. (2009). Transgenic mice expressing porcine prion protein resistant to classical scrapie but susceptible to sheep bovine spongiform encephalopathy and atypical scrapie. Emerg. Infect. Dis. *15*, 1214–1221.

Fediaevsky, A., Morignat, E., Ducrot, C., and Calavas, D. (2009). A case–control study on the origin of atypical scrapie in sheep, France. Emerg. Infect. Dis. *15*, 710–718.

Foster, J.D., Hope, J., and Fraser, H. (1993). Transmission of bovine spongiform encephalopathy to sheep and goats. Vet. Rec. *133*, 339–341.

Fraser, H., Bruce, M.E., Chree, A., McConnell, I., and Wells, G.A. (1992). Transmission of bovine spongiform encephalopathy and scrapie to mice. J. Gen. Virol. *73*, 1891–1897.

Fukuda, S., Iwamaru, Y., Imamura, M., Masujin, K., Shimizu, Y., Matsuura, Y., Shu, Y., Kurachi, M., Kasai, K., Murayama, Y., *et al.* (2009). Intraspecies transmission of L-type-like bovine spongiform encephalopathy detected in Japan. Microbiol. Immunol. *53*, 704–707.

Gambetti, P., Dong, Z., Yuan, J., Xiao, X., Zheng, M., Alshekhlee, A., Castellani, R., Cohen, M., Barria, M.A., Gonzalez-Romero, D., *et al.* (2008). A novel human disease with abnormal prion protein sensitive to protease. Ann. Neurol. *63*, 697–708.

Grassi, J., Comoy, E., Simon, S., Creminon, C., Frobert, Y., Trapmann, S., Schimmel, H., Hawkins, S.A.C., Moynagh, J., Deslys, J.R., *et al.* (2001). Rapid test for the preclinical postmortem diagnosis of BSE in central nervous system tissue. Vet. Rec. *149*, 577–582.

Gretzschel, A., Buschmann, A., Langeveld, J., and Groschup, M.H. (2006). Immunological characterization of abnormal prion protein from atypical scrapie cases in sheep using a panel of monoclonal antibodies. J. Gen. Virol. *87*, 3715–3722.

Griffiths, P.C., Spiropoulos, J., Lockey, R., Tout, A.C., Jayasena, D., Plater, J.M., Chave, A., Green, R.B., Simonini, S., Thorne, L., *et al.* (2010). Characterization of atypical scrapie cases from Great Britain in transgenic ovine PrP mice. J. Gen. Virol. *91*, 2132–2138.

Groschup, M.H., Lacroux, C., Buschmann, A., Luhken, G., Mathey, J., Eiden, M., Lugan, S., Hoffmann, C., Espinosa, J.C., Baron, T., *et al.* (2007). Classic scrapie in sheep with the ARR/ARR prion genotype in Germany and France. Emerg. Infect. Dis. *13*, 1201–1207.

Hagiwara, K., Yamakawa, Y., Sato, Y., Nakamura, Y., Tobiume, M., Shinagawa, M., and Sata, T. (2007). Accumulation of mono-glycosylated form-rich, plaque-forming PrP^{Sc} in the second atypical bovine spongiform encephalopathy case in Japan. Jpn J. Infect. Dis. *60*, 305–308.

Haritani, M., Spencer, Y.I., and Wells, G.A.H. (1994). Hydrated autoclave pretreatment enhancement of prion protein immunoreactivity in formalin-fixed bovine spongiform encephalopathy-affected brain. Acta Neuropathol. *87*, 86–90.

Hayashi, H.K., Yokoyama, T., Takata, M., Iwamaru, Y., Imamura, M., Ushiki, Y.K., and Shinagawa, M. (2005). The N-terminal cleavage site of PrP^{Sc} from BSE differs from that of PrP^{Sc} from scrapie. Biochem. Bioph. Res. Co. *328*, 1024–1027.

Hewicker-Trautwein, M., and Bradley, R. (2006). Portrait of transmissible feline spongiform encephalopathy. In Prions in Humans and Animals, Hörnlimann, B., Riesner, D., and Kretzschmar, H., eds. (Walter de Gruyter, Berlin), pp. 271–274.

Hoffmann, C., Eiden, M., Kaatz, M., Keller, M., Ziegler, U., Rogers, R., Hills, B., Balkema-Buschmann, A., van Keulen, L., Jacobs, J.G., *et al.* (2011). BSE infectivity in jejunum, ileum and ileocaecal junction of incubating cattle. Vet. Res. *42*, 21.

Hope, J., Reekie, L.J., Hunter, N., Multhaup, G., Beyreuther, K., White, H., Scott, A.C., Stack, M.J., Dawson, M., and Wells, G.A. (1988). Fibrils from brains of cows with new cattle disease contain scrapie-associated protein. Nature *336*, 390–392.

Hornlimann, B., Van Keulen, L., Ulvund, M.J., and Bradley, R. (2006). Portrait of scrapie in sheep and goat. In Prions in Humans and Animals, Hörnlimann, B., Riesner, D., and Kretzschmar, H., eds. (Walter de Gruyter, Berlin), pp. 271–274.

Ikeda, T., Horiuchi, M., Ishiguro, N., Muramatsu, Y., Kai-Uwe, G.D., and Shinagawa, M. (1995). Amino acid polymorphisms of PrP with reference to onset of scrapie in Suffolk and Corriedale sheep in Japan. J. Gen. Virol. *76*, 2577–2581.

Ikegami, Y., Ito, M., Isomura, H., Momotani, E., Sasaki, K., Muramatsu, Y., Ishiguro, N., and Shinagawa, M. (1991). Pre-clinical and clinical diagnosis of scrapie by detection of PrP protein in tissues of sheep. Vet. Rec. *128*, 271–275.

Iwamaru, Y., Imamura, M., Matsuura, Y., Masujin, K., Shimizu, Y., Shu, Y., Kurachi, M., Kasai, K., Murayama, Y., Fukuda, S., *et al.* (2010). Accumulation of L-type bovine prions in peripheral nerve tissues. Emerg. Infect. Dis. *16*, 1151–1154.

Jacobs, J.G., Langeveld, J.P., Biacabe, A.G., Acutis, P.L., Polak, M.P., Gavier-Widen, D., Buschmann, A., Caramelli, M., Casalone, C., Mazza, M., *et al.* (2007). Molecular discrimination of atypical bovine spongiform encephalopathy strains from a geographical region spanning a wide area in Europe. J. Clin. Microbiol. *45*, 1821–1829.

Jeffrey, M., Gonzalez, L., Chong, A., Foster, J., Goldmann, W., Hunter, N., and Martin, S. (2006). Ovine infection with the agents of scrapie (CH1641 isolate) and bovine spongiform encephalopathy: immunochemical similarities can be resolved by immunohistochemistry. J. Comp. Pathol. *134*, 17–29.

van Keulen, L.J., Schreuder, B.E., Meloen, R.H., Mooij-Harkes, G., Vromans, M.E., and Langeveld, J.P. (1996). Immunohistochemical detection of prion protein in lymphoid tissues of sheep with natural scrapie. J. Clin. Microbiol. *34*, 1228–1231.

Kimberlin, R.H., and Walker, C.A. (1978). Evidence that the transmission of one source of scrapie agent to hamsters involves separation of agent strains from a mixture. J. Gen. Virol. *39*, 487–496.

Kirkwood, J.K., and Cunningham, A.A. (1994). Epidemiological observations on spongiform encephalopathies in captive wild animals in the British Isles. Vet. Rec. *135*, 296–303.

Kirkwood, J.K., Cunningham, A.A., Austin, A.R., Wells, G.A.H., and Sainsbury, A.W. (1994). Spongiform encephalopathy in a greater kudu (Tragelaphus-strepsiceros) introduced into an affected group. Vet. Rec. *134*, 167–168.

Kittelberger, R., Chaplin, M.J., Simmons, M.M., Ramirez-Villaescusa, A., McIntyre, L., MacDiarmid, S.C., Hannah, M.J., Jenner, J., Bueno, R., Bayliss, D., *et al.* (2010). Atypical scrapie/Nor98 in a sheep from New Zealand. J. Vet. Diagn. Invest. *22*, 863–875.

Kong, Q., Zheng, M., Casalone, C., Qing, L., Huang, S., Chakraborty, B., Wang, P., Chen, F., Cali, I., Corona, C., *et al.* (2008). Evaluation of the human transmission risk of an atypical bovine spongiform encephalopathy prion strain. J. Virol. *82*, 3697–3701.

Konold, T., Bone, G.E., Clifford, D., Chaplin, M.J., Cawthraw, S., Stack, M.J., and Simmons, M.M. (2012). Experimental H-type and L-type bovine spongiform encephalopathy in cattle: observation of two clinical syndromes and diagnostic challenges. BMC Vet. Res. *8*, 848–854.

Lasmezas, C.I., Deslys, J.P., Demaimay, R., Adjou, K.T., Lamoury, F., Dormont, D., Robain, O., Ironside, J., and Hauw, J.J. (1996). BSE transmission to macaques. Nature *381*, 743–744.

Lee, H.S., Brown, P., Cervenakova, L., Garruto, R.M., Alpers, M.P., Gajdusek, D.C., and Goldfarb, L.G. (2001). Increased susceptibility to Kuru of carriers of the *PRNP* 129 methionine/methionine genotype. J. Infect. Dis. *183*, 192–196.

Legname, G., Nguyen, H.O., Baskakov, I.V., Cohen, F.E., Dearmond, S.J., and Prusiner, S.B. (2005). Strain-specified characteristics of mouse synthetic prions. Proc. Natl. Acad. Sci. U.S.A. *102*, 2168–2173.

Lezmi, S., Martin, S., Simon, S., Comoy, E., Bencsik, A., Deslys, J.P., Grassi, J., Jeffrey, M., and Baron, T. (2004). Comparative molecular analysis of the abnormal prion protein in field scrapie cases and experimental bovine spongiform encephalopathy in sheep by use of western blotting and immunohistochemical methods. J. Virol. *78*, 3654–3662.

Llewelyn, C.A., Hewitt, P.E., Knight, R.S.G., Amar, K., Cousens, S., Mackenzie, J., and Will, R.G. (2004). Possible transmission of variant Creutzfeldt–Jakob disease by blood transfusion. Lancet *363*, 417–421.

Lloyd, S.E., Linehan, J.M., Desbruslais, M., Joiner, S., Buckell, J., Brandner, S., Wadsworth, J.D.F., and Collinge, J. (2004). Characterization of two distinct prion strains derived from bovine spongiform encephalopathy transmissions to inbred mice. J. Gen. Virol. *85*, 2471–2478.

Lombardi, G., Casalone, C., D'Angelo, A., Gelmetti, D., Torcoli, G., Barbieri, I., Corona, C., Fasoli, E., Farinazzo, A., Fiorini, M., *et al.* (2008). Intraspecies transmission of BASE induces clinical dullness and amyotrophic changes. PLoS Pathog. *4*, e1000075.

Marsh, R.F., and Bessen, R.A. (1993). Epidemiologic and experimental studies on transmissible mink encephalopathy. Dev. Biol. Stand. *80*, 111–118.

Marsh, R.F., Bessen, R.A., Lehmann, S., and Hartsough, G.R. (1991). Epidemiological and experimental studies on a new incident of transmissible mink encephalopathy. J. Gen. Virol. *72*, 589–594.

Masujin, K., Shu, Y., Yamakawa, Y., Hagiwara, K., Sata, T., Matsuura, Y., Iwamaru, Y., Imamura, M., Okada, H., Mohri, S., *et al.* (2008). Biological and biochemical characterization of L-type-like bovine spongiform encephalopathy (BSE) detected in Japanese black beef cattle. Prion *2*, 123–128.

Masujin, K., Shu, Y.J., Okada, H., Matsuura, Y., Iwamaru, Y., Imamura, M., Mohri, S., and Yokoyama, T. (2009). Isolation of two distinct prion strains from a scrapie-affected sheep. Arch. Virol. *154*, 1929–1932.

Mazza, M., Iulini, B., Vaccari, G., Acutis, P.L., Martucci, F., Esposito, E., Peletto, S., Barocci, S., Chiappini, B., Corona, C., *et al.* (2010). Co-existence of classical scrapie and Nor98 in a sheep from an Italian outbreak. Res. Vet. Sci. *88*, 478–485.

Mestre-Frances, N., Nicot, S., Rouland, S., Biacabe, A.G., Quadrio, I., Perret-Liaudet, A., Baron, T., and Verdier, J.M. (2012). Oral transmission of L-type bovine spongiform encephalopathy in primate model. Emerg. Infect. Dis. *18*, 142–145.

Mitchell, G.B., O'Rourke, K.I., Harrington, N.P., Soutyrine, A., Simmons, M.M., Dudas, S., Zhuang, D.Y., Laude, H., and Balachandran, A. (2010). Identification of atypical scrapie in Canadian sheep. J. Vet. Diagn. Invest. *22*, 408–411.

Murayama, Y., Yoshioka, M., Masujin, K., Okada, H., Iwamaru, Y., Imamura, M., Matsuura, Y., Fukuda, S., Onoe, S., Yokoyama, T., *et al.* (2010). Sulfated dextrans enhance *in vitro* amplification of bovine spongiform encephalopathy PrP^{Sc} and enable ultrasensitive detection of bovine PrP^{Sc}. PLoS One *5*, e13152.

Nicholson, E.M., Brunelle, B.W., Richt, J.A., Kehrli, M.E., and Greenlee, J.J. (2008). Identification of a heritable polymorphism in bovine *PRNP* associated with genetic transmissible spongiform encephalopathy: Evidence of heritable BSE. PLoS One *3*, e2912.

Nicot, S., and Baron, T. (2011). Strain-specific barriers against bovine prions in hamsters. J. Virol. *85*, 1906–1908.

O'Rourke, K.I., Baszler, T.V., Besser, T.E., Miller, J.M., Cutlip, R.C., Wells, G.A., Ryder, S.J., Parish, S.M., Hamir, A.N., Cockett, N.E., *et al.* (2000). Preclinical diagnosis of scrapie by immunohistochemistry of third eyelid lymphoid tissue. J. Clin. Microbiol. *38*, 3254–3259.

Okada, H., Iwamaru, Y., Imamura, M., Masujin, K., Yokoyama, T., and Mohri, S. (2010). Immunohistochemical detection of disease-associated prion protein in the intestine of cattle naturally affected with bovine spongiform encephalopathy by using an alkaline-based chemical antigen retrieval method. J. Vet. Med. Sci. *72*, 1423–1429.

Okada, H., Iwamaru, Y., Imamura, M., Masujin, K., Matsuura, Y., Shimizu, Y., Kasai, K., Mohri, S., Yokoyama, T., and Czub, S. (2011a). Experimental H-type bovine spongiform encephalopathy characterized by plaques and glial- and stellate-type prion protein deposits. Vet. Res. *42*, 79.

Okada, H., Masujin, K., Iwamaru, Y., Imamura, M., Matsuura, Y., Mohri, S., Czub, S., and Yokoyama, T. (2011b). Experimental transmission of H-Type bovine spongiform encephalopathy to bovinized transgenic mice. Vet. Pathol. *48*, 942–947.

Okada, H., Iwamaru, Y., Kakizaki, M., Masujin, K., Imamura, M., Fukuda, S., Matsuura, Y., Shimizu, Y., Kasai, K., Mohri, S., *et al.* (2012). Properties of L-type bovine spongiform encephalopathy in intraspecies passages. Vet. Pathol. *49*, 819–823.

Ono, F., Tase, N., Kurosawa, A., Hiyaoka, A., Ohyama, A., Tezuka, Y., Wada, N., Sato, Y., Tobiume, M., Hagiwara, K., *et al.* (2011). Atypical L-type bovine spongiform encephalopathy (L-BSE) transmission to cynomolgus macaques, a non-human primate. Jpn J. Infect. Dis. *64*, 81–84.

Owen, J.P., Rees, H.C., Maddison, B.C., Terry, L.A., Thorne, L., Jackman, R., Whitelam, G.C., and Gough, K.C. (2007). Molecular profiling of ovine prion diseases by using thermolysin-resistant PrP^{Sc} and endogenous C2 PrP fragments. J. Virol. *81*, 10532–10539.

Prusiner, S.B. (1991). Molecular biology of prion diseases. Science *252*, 1515–1522.

Race, R., Ernst, D., Jenny, A., Taylor, W., Sutton, D., and Caughey, B. (1992). Diagnostic implications of detection of proteinase K-resistant protein in spleen, lymph nodes, and brain of sheep. Am. J. Vet. Res. *53*, 883–889.

Richt, J.A., Kunkle, R.A., Alt, D., Nicholson, E.M., Hamir, A.N., Czub, S., Kluge, J., Davis, A.J., and Hall, S.M. (2007). Identification and characterization of two bovine spongiform encephalopathy cases diagnosed in the United States. J. Vet. Diagn. Invest. *19*, 142–154.

Robinson, M.M., Hadlow, W.J., Huff, T.P., Wells, G.A., Dawson, M., Marsh, R.F., and Gorham, J.R. (1994). Experimental infection of mink with bovine spongiform encephalopathy. J. Gen. Virol. *75*, 2151–2155.

Schaller, O., Fatzer, R., Stack, M., Clark, J., Cooley, W., Biffiger, K., Egli, S., Doherr, M., Vandevelde, M., Heim, D., *et al.* (1999). Validation of a Western immunoblotting procedure for bovine PrP^{Sc} detection and its use as a rapid surveillance method for the diagnosis of bovine spongiform encephalopathy (BSE). Acta Neuropathol. *98*, 437–443.

Schreuder, B.E., van Keulen, L.J., Vromans, M.E., Langeveld, J.P., and Smits, M.A. (1996). Preclinical test for prion diseases. Nature *381*, 563.

Scott, M.R., Will, R., Ironside, J., Nguyen, H.O.B., Tremblay, P., DeArmond, S.J., and Prusiner, S.B. (1999). Compelling transgenetic evidence for transmission of bovine spongiform encephalopathy prions to humans. Proc. Natl. Acad. Sci. U.S.A. *96*, 15137–15142.

Seuberlich, T., Gsponer, M., Drogemuller, C., Polak, M.P., McCutcheon, S., Heim, D., Oevermann, A., and Zurbriggen, A. (2012). Novel prion protein in BSE-affected cattle, Switzerland. Emerg. Infect. Dis. *18*, 158–159.

Shimada, K., Hayashi, H.K., Ookubo, Y., Iwamaru, Y., Imamura, M., Takata, M., Schmerr, M.J., Shinagawa, M., and Yokoyama, T. (2005). Rapid PrP^{Sc} detection in lymphoid tissue and application to scrapie surveillance of fallen stock in Japan: variable PrP^{Sc} accumulation in palatal tonsil in natural scrapie. Microbiol. Immunol. *49*, 801–804.

Shu, Y.J., Masujin, K., Okada, H., Iwamaru, Y., Imamura, M., Matsuura, Y., Mohri, S., and Yokoyama, T. (2011). Characterization of Syrian hamster adapted prions derived from L-type and C-type bovine spongiform encephalopathies. Prion *5*, 103–109.

Simmons, M.M., Konold, T., Simmons, H.A., Spencer, Y.I., Lockey, R., Spiropoulos, J., Everitt, S., and Clifford, D. (2007). Experimental transmission of atypical scrapie to sheep. BMC Vet. Res. *3*, 20.

Simmons, M.M., Konold, T., Thurston, L., Bellworthy, S.J., Chaplin, M.J., and Moore, S.J. (2010). The natural atypical scrapie phenotype is preserved on experimental transmission and sub-passage in *PRNP* homologous sheep. BMC Vet. Res. *6*, 14.

Simmons, M.M., Moore, S.J., Konold, T., Thurston, L., Terry, L.A., Thorne, L., Lockey, R., Vickery, C., Hawkins, S.A., Chaplin, M.J., *et al.* (2011). Experimental oral transmission of atypical scrapie to sheep. Emerg. Infect. Dis. *17*, 848–854.

Siso, S., Jeffrey, M., Houston, F., Hunter, N., Martin, S., and Gonzalez, L. (2010). Pathological phenotype of sheep scrapie after blood transfusion. J. Comp. Pathol. *142*, 27–35.

Stack, M., Jeffrey, M., Gubbins, S., Grimmer, S., Gonzalez, L., Martin, S., Chaplin, M., Webb, P., Simmons, M., Spencer, Y., *et al.* (2006). Monitoring for bovine spongiform encephalopathy in sheep in Great Britain, 1998–2004. J. Gen. Virol. *87*, 2099–2107.

Stack, M.J., Moore, S.J., Davis, A., Webb, P.R., Bradshaw, J.M., Lee, Y.H., Chaplin, M., Focosi-Snyman, R., Thurston, L., Spencer, Y.I., *et al.* (2011a). Bovine spongiform encephalopathy: investigation of phenotypic variation among passive surveillance cases. J. Comp. Pathol. *144*, 277–288.

Stack, M.J., Moore, S.J., Vidal-Diez, A., Arnold, M.E., Jones, E.M., Spencer, Y.I., Webb, P., Spiropoulos, J., Powell, L., Bellerby, P., *et al.* (2011b). Experimental bovine spongiform encephalopathy: detection of PrP^{Sc} in the small intestine relative to exposure dose and age. J. Comp. Pathol. *145*, 289–301.

Telling, G.C., Parchi, P., DeArmond, S.J., Cortelli, P., Montagna, P., Gabizon, R., Mastrianni, J., Lugaresi, E., Gambetti, P., and Prusiner, S.B. (1996). Evidence for the conformation of the pathologic isoform of the prion protein enciphering and propagating prion diversity. Science *274*, 2079–2082.

Terry, L.A., Jenkins, R., Thorne, L., Everest, S.J., Chaplin, M.J., Davis, L.A., and Stack, M.J. (2007). First case of H-type bovine spongiform encephalopathy identified in Great Britain. Vet. Rec. *160*, 873–874.

Tester, S., Juillerat, V., Doherr, M.G., Haase, B., Polak, M., Ehrensperger, F., Leeb, T., Zurbriggen, A., and Seuberlich, T. (2009). Biochemical typing of pathological prion protein in aging cattle with BSE. Virol. J. *6*, 64.

Torres, J.M., Andreoletti, O., Lacroux, C., Prieto, I., Lorenzo, P., Larska, M., Baron, T., and Espinosa, J.C. (2011). Classical bovine spongiform encephalopathy by transmission of H-type prion in homologous prion protein context. Emerg. Infect. Dis. *17*, 1636–1644.

Ushiki-Kaku, Y., Endo, R., Iwamaru, Y., Shimizu, Y., Imamura, M., Masujin, K., Yamamoto, T., Hattori, S., Itohara, S., Irie, S., *et al.* (2010). Tracing conformational transition of abnormal prion proteins during interspecies transmission by using novel antibodies. J. Biol. Chem. *285*, 11931–11936.

Wells, G.A.H., Scott, A.C., Johnson, C.T., Gunning, R.F., Hancock, R.D., Jeffrey, M., Dawson, M., and Bradley, R. (1987). A novel progressive spongiform encephalopathy in cattle. Vet. Rec. *121*, 419–420.

Westaway, D., Goodman, P.A., Mirenda, C.A., McKinley, M.P., Carlson, G.A., and Prusiner, S.B. (1987). Distinct prion proteins in short and long scrapie incubation period mice. Cell *51*, 651–662.

Westaway, D., Zuliani, V., Cooper, C.M., Da Costa, M., Neuman, S., Jenny, A.L., Detwiler, L., and Prusiner, S.B. (1994). Homozygosity for prion protein alleles encoding glutamine-171 renders sheep susceptible to natural scrapie. Genes Dev. *8*, 959–969.

Wilesmith, J.W., Wells, G.A.H., Cranwell, M.P., and Ryan, J.B.M. (1988). Bovine spongiform encephalopathy – epidemiological studies. Vet. Rec. *123*, 638–644.

Yamakawa, Y., Hagiwara, K., Nohtomi, K., Nakamura, Y., Nishijima, M., Higuchi, Y., Sato, Y., and Sata, T. (2003). Atypical proteinase K-resistant prion protein (PrP^{res}) observed in an apparently healthy 23-month-old Holstein steer. Jpn J. Infect. Dis. *56*, 221–222.

Yokoyama, T., Masujin, K., Iwamaru, Y., Imamura, M., and Mohri, S. (2009). Alteration of the biological and biochemical characteristics of bovine spongiform encephalopathy prions during interspecies transmission in transgenic mice models. J. Gen. Virol. *90*, 261–268.

Yokoyama, T., Masujin, K., Schmerr, M.J., Shu, Y.J., Okada, H., Iwamaru, Y., Imamura, M., Matsuura, Y., Murayama, Y., and Mohri, S. (2010). Intraspecies prion transmission results in selection of sheep scrapie strains. PLoS One *5*, e15450.

Chronic Wasting Disease and Other Animal Prion Diseases

10

Akikazu Sakudo

Abstract

Chronic wasting disease (CWD) is one of the prion diseases showing clinical symptoms such as progressive weight loss, abnormal behaviour and excessive salivation. Incidents have been reported in North America and Korea. Efficient transmission by horizontal infection possibly occurs via saliva or faeces. Infectivity in the skeletal muscle of infected deer has been observed, suggesting meat could be a potential source of infection. However, at the moment, CWD transmission to humans is thought to be very unlikely. In this chapter, we also introduce representatives of other animal prion diseases, such as transmissible mink encephalopathy (TME) and feline spongiform encephalopathy (FSE). Recently, possible transmission of transmissible spongiform encephalopathy (TSE) agent to fishes has also been reported.

History of chronic wasting disease

Chronic wasting disease (CWD) is a prion disease transmitted in deer such as mule deer (*Odocoileus hemionus*), white-tailed deer (*Odocoileus virginianus*), Rocky Mountain elk (*Cervus elaphus nelsoni*) and Shira's moose (*Alces alces shirasi*) (Williams *et al.*, 1980, 1982; Baeten *et al.*, 2007). CWD has extraordinary features of incidence in wild animals.

The first report of CWD described a wasting disorder in mule deer caught for nutrition research near Fort Collins, Colorado in 1967 (Williams *et al.*, 1980). In 1978, pathologists Elizabeth Williams and Stewart Young found that CWD is a form of spongiform encephalopathy (Williams *et al.*, 1980). It was subsequently discovered that CWD results not only in typical vacuolation in neurons (Williams *et al.*, 1980) but also in prion protein (PrP) accumulation (Spraker *et al.*, 2002) and infectivity in the brain (Browning *et al.*, 2004).

In 1981, a captive CWD-infected elk was found in Colorado. In 1985, a captive CWD-infected mule deer was also identified. As a result of increased surveillance, CWD was also discovered in Wyoming, South Dakota, Utah, Wisconsin, Illinois, Nebraska and New Mexico. Indeed, more than 30 million deer are thought to be infected with CWD in North America. CWD is thought to be horizontally infected at a high rate among cervids (Miller *et al.*, 2003). Oral transmission of CWD-infected brain homogenate derived from mule deer to elk has been reported, which is further supported by the finding of CWD-infected free-ranging moose (Baeten *et al.*, 2007). Although the incidence of CWD is limited to North America and Korea (Kim *et al.*, 2005; Williams, 2005; Dube *et al.*, 2006), the origins of the disease remain unclear. Prevalence rates of infection as high as 80% have been reported among does and yearling white-tailed deer in farms and as high as 15% in free-ranging

Table 10.1 Animals susceptible to CWD

Species	References
Infection in wildlife	
Mule deer (*Odocoileus hemionus*)	Williams *et al*. (1980)
White-tailed deer (*Odocoileus virginianus*)	Williams *et al*. (2002)
Rocky Mountain elk (*Cervus elaphus nelson*)	Williams *et al*. (1982)
Moose (*Alces alces shirasi*)	Kreeger *et al*. (2006)
Goat	Williams *et al*. (1992)
Experimental infection in laboratory	
Ferret	Bartz *et al*. (1998), Sigurdson *et al*. (2008)
Raccoon	Hamir *et al*. (2003)
Squirrel monkey	Marsh *et al*. (2005)
Bovine	Hamir *et al*. (2005)
Sheep	Williams (2005)
Meadow voles	Heisey *et al*. (2010)
Red backed voles	Heisey *et al*. (2010)
White-footed mice	Heisey *et al*. (2010)
Bank voles	Heisey *et al*. (2010)
Deer mice	Heisey *et al*. (2010)

white-tailed bucks in the core endemic area of CWD in southern Wisconsin, where the occurrence of this disease is typically much lower (Kaene *et al.*, 2008). Saprovores, such as buzzard and jackal, and other animals, such as mountain lion, fox, raccoon, coyote and eagle, prey on deer and elk. Thus, CWD may potentially spread widely among a variety of animals. So far, CWD-susceptible animals in addition to cervids include ferret (Bartz *et al.*, 1998), raccoon (Hamir *et al.*, 2003), squirrel monkey (Marsh *et al.*, 2005), cattle (Hamir *et al.*, 2005), sheep (Williams, 2005), goat (Williams *et al.*, 1992), hamsters (Bartz *et al.*, 1998; Raymond *et al.*, 2007), bank voles (Heisey *et al.*, 2010), mink (Harrington *et al.*, 2008), meadow voles (Heisey *et al.*, 2010), red backed voles (Heisey *et al.*, 2010), white-footed mice (Heisey *et al.*, 2010) and deer mice (Heisey *et al.*, 2010) (Table 10.1).

The concentration of prions in seawater and rivers and their stability in these environments remain unclear. It has been demonstrated that hamster-derived PrP^{Sc} can adhere to soil minerals (Johnson *et al.*, 2006). In addition, binding to soil enhances oral transmissibility of hamster adapted mink prions (Johnson *et al.*, 2007). These findings suggest that soil may be a source of CWD infection. Taken together, these observations emphasize the importance of minimizing environmental pollution, particularly in soil and water.

Clinical symptoms and pathogenesis of CWD

After infection with CWD agent, cellular prion protein (PrP^{C}) is converted into proteinase K (PK)-resistant PrP (i.e. an abnormal isoform of cervid PrP, PrP^{CWD}). PrP^{CWD} accumulates

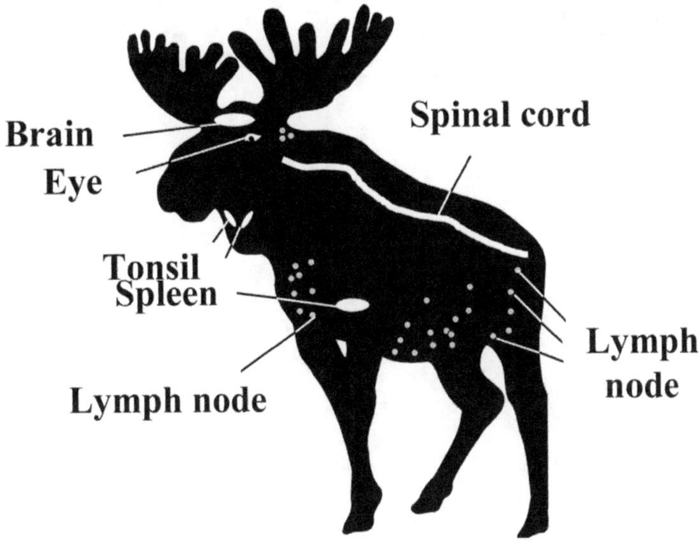

Figure 10.1 The distribution of PrP^{CWD}.

in brain tissue causing neurological symptoms and ultimately death. The distribution of PrP^{CWD} is primarily restricted to the central nervous system (CNS) and lymphatic tissues. These include brain, spinal cord, tonsil, lymph node and spleen (Fig. 10.1). Small quantities of PrP^{CWD} agent are also found in the heart and eyes of infected deer. CWD prions in saliva (Haley *et al.*, 2009), urine (Haley *et al.*, 2009), faeces (Tamgüney *et al.*, 2009) and blood (Mathiason *et al.*, 2006) have also been detected. For example, orally infected CWD-brain homogenate of mule deer induce PrP^{CWD} accumulation in intestine-associated lymphatic tissues, such as retropharyngeal lymph nodes, tonsil, Peyer's patch and ileocecal lymph nodes, as early as 42 days post-infection. Therefore, it has been proposed that CWD agent is transmitted via saliva or in the stools of deer (Miller *et al.*, 2004). More importantly, skeletal muscle of CWD-infected deer causes CWD in cervid PrP expressing transgenic mice (Angers *et al.*, 2006). This finding indicates that muscle is a potential source of infection, suggesting that the meat of deer must to be treated with caution. The characteristic clinical symptoms of CWD-infected animals are weight loss, emaciation, excessive salivation, teeth grinding, fever, anorexia, polyposia, excessive urination, impaired motor coordination and respiratory distress (Table 10.2) (Sohn *et al.*, 2002; Williams, 2005). Because these symptoms are not CWD specific, additional biochemical and pathological analysis is required to verify the diagnosis. Although CWD-infected deer showing symptoms of the disease can be found in all four seasons, infected carcasses are most common in the winter. This is presumably due to the severe winter climate in North America. Furthermore, most deer displaying symptoms of CWD are 3–4 years of age.

Unfortunately, it remains unclear how CWD is spread. Direct contact between animals and indirect transmission via soil and body surfaces are potential mechanisms. Transfer of infection in saliva and stools are currently considered a likely means of transmission. For example, a carcass put into a field successfully transmitted CWD to experimental mule deer (Miller *et al.*, 2004), suggesting a contaminated CWD-environment can result in the spread

Table 10.2 Symptoms of CWD

Weight loss
Excessive salivation
Teeth grinding
Fever
Respiratory distress
Head tremors
Ataxia
Difficulty in swallowing
Aspiration pneumonia
Depression
Isolation from herd
Hyper excitability
Regurgitation
Polyuria/polydipsia
No awareness of surroundings

of CWD. Evidence for horizontal transmission of CWD has accumulated from epidemiological data as well as experimental model experiments using transgenic mice expressing the normal cervid prion protein (Seelig *et al.*, 2010). Although several transmissible spongiform encephalopathies, such as scrapie and bovine spongiform encephalopathy (BSE), show similar symptoms, there are no relationships between such transmissible spongiform encephalopathies (TSEs) and CWD. In addition, Centers for Disease Control and Prevention (CDC) report that the possibility of transmission to humans is very low, concluding that there is little relationship between CWD and human prion diseases. This conclusion is supported by the fact that transgenic mice expressing human PrP did not show any hallmark symptoms of CWD 600 days after intracerebral inoculation with CWD prions (Kong *et al.*, 2005). These findings indicate the presence of a barrier that makes interspecies transmission unlikely, especially between species as diverse as humans and deer (Kong *et al.*, 2005). By comparison, the incubation time, which indicates the period required for displaying symptoms after intracerebral infection with CWD prions, using transgenic mice that had been 'cervidized' was between 118–142 days (Kong *et al.*, 2005). Moreover, an *in vitro* assay showed that PrP^{CWD}-mediated conversion of human PrP to a protease-resistant form is very inefficient, suggesting a species barrier at the molecular level that limits the susceptibility of humans to CWD (Raymond *et al.*, 2000). Nonetheless, given that the information on CWD is limited, the consumption of CWD-positive animals should be prohibited at this time.

Diagnostic methods for CWD

The gold standard of CWD diagnosis is immunohistochemistry (IHC). Medullary obex and lymphatic tissues, such as tonsil and pharyngeal lymph nodes, are used for the analysis.

IHC is a method of observing tissue sections stained with anti-PrP antibody using light microscopy. This methodology has merits in terms of its high level of sensitivity and specificity. In addition, IHC can be used to verify whether the tissue distribution of PrP^{CWD} is as anticipated. Other methods for CWD diagnosis include enzyme-linked immunosorbent assay (ELISA) and Western blotting. ELISA is used to detect PrP adsorbed on plates after treatment of brain homogenate with proteinase K (PK). Western blotting detects PrP absorbed on a membrane after an electroblotting procedure. Assuming that the sample is not is short supply, ELISA is preferable because the technique allows rapid analysis. However, Western blotting facilitates more detailed biochemical analysis such as determination of the molecular weight of PrP. Such information may be used to help identify the prion source and strains. Thus far, Western blot analysis of glycoform patterns of PrP^{CWD} suggest that CWD in deer and elk must be considered a single disease (Race *et al.*, 2002). However, recent studies indicate the presence of conformational variants or strains may exist. This is proposed by the evidence that sequential passage of a single mule deer CWD isolate using hamster or hamster-PrP expressing transgenic mice causes two distinct disease phenotypes (Raymond *et al.*, 2007; Heisey *et al.*, 2010). In addition, LaFauci *et al.* have reported that elk PrP-expressing transgenic mice developed phenotypically divergent diseases when inoculated with either mule deer or elk CWD, which was suggestive of different strains among cervids (LaFauci *et al.*, 2006). An ELISA kit has been certified as an official method of testing for CWD by the United States Department of Agriculture (USDA). Recent developments in detecting CWD have enabled antemortem diagnosis using IHC of tonsil tissue. More recently, a new method for prion diagnosis has been developed known as protein misfolding cyclic amplification (PMCA). This method can amplify PK-resistant PrP by sequential incubation and sonication using a mixture of PrP^{C} source and PrP^{Sc} seed. PMCA can detect PrP^{CWD} in urine of presymptomatic deer (Rubenstein *et al.*, 2011). Recently, Hoover *et al.* successfully amplified PrP^{CWD} using brain homogenate of non-cervids (ferret and hamster) (Kurt *et al.*, 2009). Using this method, these workers attempted to amplify PrP^{CWD} using brain homogenate derived from various rodents. The results indicated successful amplification using PrPs derived from bank vole and field mouse but failed using those from prairie dog and coyote (Kurt *et al.*, 2009). In addition, this method can amplify PK-resistant PrP derived not only from cervids (Kurt *et al.*, 2007), but also various species including hamster, mouse, sheep and human (Seelig *et al.*, 2010). Moreover, a transgenic mouse expressing cervid PrP (Browning *et al.*, 2004; Seelig *et al.*, 2010) and a CWD-susceptible cell line (Raymond *et al.*, 2006) have been developed. These powerful research tools will be invaluable to CWD research in helping to reveal the molecular basis of the disease. In particular, these tools will be used to develop bioassays for CWD infectivity in tissues and fluid from CWD-infected cervids (Angers *et al.*, 2006). Such an assay will contribute to risk assessment of deer tissue obtained from CWD-endemic areas.

Transmissible mink encephalopathy (TME) and other animal prion diseases

Transmissible mink encephalopathy (TME) is a prion disease of mink (Liberski *et al.*, 2009). After infection of TME agent the animal displays abnormal behaviour in the early stage, followed by increased levels of aggression and eventually hypersensitive reaction to contact and noise in the latter stage. Finally, the animal becomes immobile, resulting in gradual debilitation and death. Experiments show that the TME agent can be transmitted by subcutaneous and intradermal injection but not by oral injection. These findings suggest TME infection occurs via a wound. The TME agent was found to infect hamster, ferret, sheep, cattle and squirrel monkey, but not mouse. Animals infected with TSE besides BSE, scrapie, CWD and TME are cat, nyala, gemsbok (*Oryx gazella*), Arabian oryx (*Oryx leucoryx*), kudu (*Tragelaphus strepsiceros*), eland (*Taurotragus oryx* or *Taurotragus derbianus*), mouflon (*Ovis musimon*), cougar (*Puma concolor*), cheetah (*Acinonyx jubatus*), tiger (*Panthera tigris*) and white oryx (*Oryx dammah*) (Harris, 2010).

It should be noted that recent TSE infection experiments using fish have been reported. Oral injection of scrapie prion to rainbow trout (*Oncorhynchus mykiss*) or turbot (*Scophthalmus maximus*) showed that there was no infectivity in any of the tissues, including brain and intestine, from 15 to 90 days post infection. Therefore, prions are thought to be cleared rapidly without being absorbed through the intestinal tract of fishes (Ingrosso *et al.*, 2006). Another study showed that oral injection of BSE and scrapie prion into gilt-head bream (*Sparus aurata*) caused an accumulation of PK-resistant PrP in their brains after a period of 2 years (Salta *et al.*, 2009). However, there are no results from animal bioassay for examining the infectious titre of the prions. Thus, further studies are required to investigate the mode of infection.

Acknowledgement

This work was supported by Grant-in-Aids for Scientific Research from the Ministry of Education, Science, Culture and Technology of Japan (23780299), Grant-in-Aid for Promotion of Basic Research Activities for Innovative Biosciences from Bio-oriented Technology Research Advancement Institution (BRAIN), Grants-in-Aid from the Research Committee of Prion disease and Slow Virus Infection, the Ministry of Health, Labour and Welfare of Japan, and Grant-in-Aid for Government and Administration related research from Senri Life Science Foundation.

References

Angers, R.C., Browning, S.R., Seward, T.S., Sigurdson, C.J., Miller, M.W., Hoover, E.A., and Telling, G.C. (2006). Prions in skeletal muscles of deer with chronic wasting disease. Science *311*, 1117.

Baeten, L.A., Powers, B.E., Jewell, J.E., Spraker, T.R., and Miller, M.W. (2007). A natural case of chronic wasting disease in a free-ranging moose (Alces alces shirasi). J. Wildl. Dis. *43*, 309–314.

Bartz, J.C., Marsh, R.F., McKenzie, D.I., and Aiken, J.M. (1998). The host range of chronic wasting disease is altered on passage in ferrets. Virology *251*, 297–301.

Browning, S.R., Mason, G.L., Seward, T., Green, M., Eliason, G.A., Mathiason, C., Miller, M.W., Williams, E.S., Hoover, E., and Telling, G.C. (2004). Transmission of prions from mule deer and elk with chronic wasting disease to transgenic mice expressing cervid PrP. J. Virol. *78*, 13345–13350.

Dubé, C., Mehren, K.G., Barker, I.K., Peart, B.L., and Balachandran, A. (2006). Retrospective investigation of chronic wasting disease of cervids at the Toronto Zoo, 1973–2003. Can. Vet. J. *47*, 1185–1193.

Fernández-Borges, N., de Castro, J., and Castilla, J. (2009). *In vitro* studies of the transmission barrier. Prion 3, 220–223.

Haley, N.J., Seelig, D.M., Zabel, M.D., Telling, G.C., and Hoover, E.A. (2009). Detection of CWD prions in urine and saliva of deer by transgenic mouse bioassay. PLoS One *4*, e4848.

Hamir, A.N., Miller, J.M., Cutlip, R.C., Stack, M.J., Chaplin, M.J., Jenny, A.L., and Williams, E.S. (2003). Experimental inoculation of scrapie and chronic wasting disease agents in raccoons (Procyon lotor). Vet Rec. *153*, 121–123.

Hamir, A.N., Kunkle, R.A., Cutlip, R.C., Miller, J.M., O'Rourke, K.I., Williams, E.S., Miller, M.W., Stack, M.J., Chaplin, M.J., and Richt, J.A. (2005). Experimental transmission of chronic wasting disease agent from mule deer to cattle by the intracerebral route. J. Vet. Diagn. Invest. *17*, 276–281.

Harrington, R.D., Baszler, T.V., O'Rourke, K.I., Schneider, D.A., Spraker, T.R., Liggitt, H.D., and Knowles, D.P. (2008). A species barrier limits transmission of chronic wasting disease to mink (*Mustela vison*). J. Gen. Virol. *89*, 1086–1096.

Harris, D., ed. (2010). Mad Cow Disease and Related Spongiform Encephalopathies (Current Topics in Microbiology and Immunology) (Springer, New York).

Heisey, D.M., Mickelsen, N.A., Schneider, J.R., Johnson, C.J., Langenberg, J.A., Bochsler, P.N., Keane, D.P., and Barr, D.J. (2010). Chronic wasting disease (CWD) susceptibility of several North American rodents that are sympatric with cervid CWD epidemics. J. Virol. *84*, 210–215.

Ingrosso, L., Novoa, B., Valle, A.Z., Cardone, F., Aranguren, R., Sbriccoli, M., Bevivino, S., Iriti, M., Liu, Q., Vetrugno, V., *et al.* (2006). Scrapie infectivity is quickly cleared in tissues of orally-infected farmed fish. BMC Vet. Res. *2*, 21.

Johnson, C.J., Phillips, K.E., Schramm, P.T., McKenzie, D., Aiken, J.M., and Pedersen, J.A. (2006). Prions adhere to soil minerals and remain infectious. PLoS Pathog. *2*, e32.

Johnson, C.J., Pedersen, J.A., Chappell, R.J., McKenzie, D., and Aiken, J.M. (2007). Oral transmissibility of prion disease is enhanced by binding to soil particles. PLoS Pathog. *3*, e93.

Keane, D.P., Barr, D.J., Bochsler, P.N., Hall, S.M., Gidlewski, T., O'Rourke, K.I., Spraker, T.R., and Samuel, M.D. (2008). Chronic wasting disease in a Wisconsin white-tailed deer farm. J. Vet. Diagn. Invest. *20*, 698–703.

Kim, T.Y., Shon, H.J., Joo, Y.S., Mun, U.K., Kang, K.S., and Lee, Y.S. (2005). Additional cases of Chronic Wasting Disease in imported deer in Korea. J. Vet. Med. Sci. *67*, 753–759.

Kong, Q., Huang, S., Zou, W., Vanegas, D., Wang, M., Wu, D., Yuan, J., Zheng, M., Bai, H., Deng, H., *et al.* (2005). Chronic wasting disease of elk: transmissibility to humans examined by transgenic mouse models. J. Neurosci. *25*, 7944–7949.

Kreeger, T.J., Montgomery, D.L., Jewell, J.E., Schultz, W., and Williams, E.S. (2006). Oral transmission of chronic wasting disease in captive Shira's moose. J. Windl. Dis. *42*, 640–645.

Kurt, T.D., Perrott, M.R., Wilusz, C.J., Wilusz, J., Supattapone, S., Telling, G.C., Zabel, M.D., and Hoover, E.A. (2007). Efficient *in vitro* amplification of chronic wasting disease PrPRES. J. Virol. *81*, 9605–9608.

Kurt, T.D., Telling, G.C., Zabel, M.D., and Hoover, E.A. (2009). Trans-species amplification of PrP(CWD) and correlation with rigid loop 170N. Virology *387*, 235–243.

LaFauci, G., Carp, R.I., Meeker, H.C., Ye, X., Kim, J.I., Natelli, M., Cedeno, M., Petersen, R.B., Kascsak, R., and Rubenstein, R. (2006). Passage of chronic wasting disease prion into transgenic mice expressing Rocky Mountain elk (*Cervus elaphus nelsoni*) PrPC. J. Gen. Virol. *87*, 3773–3780.

Liberski, P.P., Sikorska, B., Guiroy, D., and Bessen, R.A. (2009).Transmissible mink encephalopathy – review of the etiology of a rare prion disease. Folia Neuropathol. *47*, 195–204.

Marsh, R.F., Kincaid, A.E., Bessen, R.A., and Bartz, J.C. (2005). Interspecies transmission of chronic wasting disease prions to squirrel monkeys (*Saimiri sciureus*). J Virol. *79*, 13794–13796.

Mathiason, C.K., Powers, J.G., Dahmes, S.J., Osborn, D.A., Miller, K.V., Warren, R.J., Mason, G.L., Hays, S.A., Hayes-Klug, J., Seelig, D.M., *et al.* (2006). Infectious prions in the saliva and blood of deer with chronic wasting disease. Science *314*, 133–136.

Miller, M.W., and Williams, E.S. (2003). Prion disease: horizontal prion transmission in mule deer. Nature *425*, 35–36.

Miller, M.W., Williams, E.S., Hobbs, N.T., and Wolfe, L.L. (2004). Environmental sources of prion transmission in mule deer. Emerg. Infect. Dis. *10*, 1003–1006.

Race, R.E., Raines, A., Baron, T.G., Miller, M.W., Jenny, A., and Williams, E.S. (2002). Comparison of abnormal prion protein glycoform patterns from transmissible spongiform encephalopathy agent-infected deer, elk, sheep, and cattle. J. Virol. *76*, 12365–12368.

Race, B., Meade-White, K.D., Miller, M.W., Barbian, K.D., Rubenstein, R., LaFauci, G., Cervenakova, L., Favara, C., Gardner, D., Long, D., *et al.* (2009). Susceptibilities of nonhuman primates to chronic wasting disease. Emerg. Infect. Dis. *15*, 1366–1376.

Raymond, G.J., Bossers, A., Raymond, L.D., O'Rourke, K.I., McHolland, L.E., Bryant, P.K. 3rd., Miller, M.W., Williams, E.S., Smits, M., and Caughey, B. (2000). Evidence of a molecular barrier limiting susceptibility of humans, cattle and sheep to chronic wasting disease. EMBO J. *19*, 4425–4430.

Raymond, G.J., Olsen, E.A., Lee, K.S., Raymond, L.D., Bryant, P.K. 3rd., Baron, G.S., Caughey, W.S., Kocisko, D.A., McHolland, L.E., Favara, C., *et al.* (2006). Inhibition of protease-resistant prion protein formation in a transformed deer cell line infected with chronic wasting disease. J. Virol. *80*, 596–604.

Raymond, G.J., Raymond, L.D., Meade-White, K.D., Hughson, A.G., Favara, C., Gardner, D., Williams, E.S., Miller, M.W., Race, R.E., and Caughey, B. (2007). Transmission and adaptation of chronic wasting disease to hamsters and transgenic mice: evidence for strains. J. Virol. *81*, 4305–4314.

Rubenstein, R., Chang, B., Gray, P., Piltch, M., Bulgin, M.S., Sorensen-Melson, S., and Miller, M.W. (2011). Prion disease detection, PMCA kinetics, and IgG in urine from sheep naturally/experimentally infected with scrapie and deer with preclinical/clinical chronic wasting disease. J. Virol. *85*, 9031–9038.

Salta, E., Panagiotidis, C., Teliousis, K., Petrakis, S., Eleftheriadis, E., Arapoglou, F., Grigoriadis, N., Nicolaou, A., Kaldrymidou, E., Krey, G., *et al.* (2009). Evaluation of the possible transmission of BSE and scrapie to gilthead sea bream (*Sparus aurata*). PLoS One *4*, e6175.

Seelig, D.M., Mason, G.L., Telling, G.C., and Hoover, E.A. (2010).Pathogenesis of chronic wasting disease in Cervidized transgenic mice. Am. J. Pathol. *176*, 2785–2797.

Sigurdson, C.J., Mathiason, C.K., Perrott, M.R., Eliason, G.A., Spraker, T.R., Glatzel, M., Manco, G., Bartz, J.C., Miller, M.W., and Hoover, E.A. (2008). Experimental chronic wasting disease (CWD) in the ferret. J. Comp. Pathol. *138*, 189–196.

Sohn, H.J., Kim, J.H., Choi, K.S., Nah, J.J., Joo, Y.S., Jean, Y.H., Ahn, S.W., Kim, O.K., Kim, D.Y., and Balachandran, A. (2002). A case of chronic wasting disease in an elk imported to Korea from Canada. J. Vet. Med. Sci. *64*, 855–858.

Spraker, T.R., Zink, R.R., Cummings, B.A., Wild, M.A., Miller, M.W., and O'Rourke, K.I. (2002). Comparison of histological lesions and immunohistochemical staining of proteinase-resistant prion protein in a naturally occurring spongiform encephalopathy of free-ranging mule deer (Odocoileus hemionus) with those of chronic wasting disease of captive mule deer. Vet Pathol. *39*, 110–119.

Tamgüney, G., Miller, M.W., Wolfe, L.L., Sirochman, T.M., Glidden, D.V., Palmer, C., Lemus, A., DeArmond, S.J., and Prusiner, S.B. (2009). Asymptomatic deer excrete infectious prions in faeces. Nature *461*, 529–532.

Williams, E.S. (2005). Chronic wasting disease. Vet. Pathol. *42*, 530–549.

Willams, E.S., and Miller, M.W., (2002). Chronic wasting disease in deer and elk in North America. Am. Rev. Sci. Tech. *21*, 305–316.

Williams, E.S., and Young, S. (1980). Chronic wasting disease of captive mule deer: a spongiform encephalopathy. J. Wildl. Dis. *16*, 89–98.

Williams, E.S., and Young, S. (1982). Spongiform encephalopathy of Rocky Mountain elk. J. Windl. Dis. *18*, 465–471.

Williams, E.S., and Young, S. (1992). Spongiform encephalopathies in Cervidae. Rev. Sci. Tech. *11*, 551–567.

Future Prospects

11

Takashi Onodera and Katsuaki Sugiura

Abstract

Results from the Netherlands show that classic scrapie control can be obtained at the national scale without a loss of genetic polymorphism from any of sheep breed. No classical scrapie strain thus far has escaped ARR-associated resistance. Ongoing studies show that atypical scrapie strain also was controlled by ARR-associated resistance. In line with this expectation, the breeding programme proved successful in Dutch flock in 2010. When considering the rapid outbreak control as observed in the Netherlands study, the use of resistant rams seems sufficient and can be recommended as a control strategy in scrapie-affected countries.

The origin of atypical BSE cases is currently unknown. As with classical BSE, exposure of these animals to feed contaminated with low titres of TSE agent cannot be excluded, although other origins for these TSE forms cannot be discarded. In particular, the unusually old ages of all H-BSE and L-BSE identified cases and their apparent low-prevalence in the population could suggest that these atypical BSE forms are arising spontaneously.

PMCA needs to be highly standardized and robust in terms of a consistent and objectively quantifiable PrP^{res} amplification if to be used for quantification of the proteinaceous seeding activity of prions. There is a direct quantitative correspondence between the seeding and infectious activities of 263K scrapie prions measured by RT-QuIC and bioassay. The methodological, conceptual and practical results described in the report of 263K scrapie prions should be validated for the most human TSE agents.

Introduction

Although scrapie has been known for decades, relatively little attention has been paid to it as a natural disease of sheep and goats mainly because the economical impact has been relatively small compared with other diseases in sheep. The occurrence of bovine spongiform encephalopathy (BSE) provide a new impetus to research into the transmissible spongiform encephalopathies (TSE) (Wells *et al.*, 1987). Not only was the economical impact of BSE much greater than that of scrapie, the link with variant Creutzfeldt–Jakob disease (vCJD) in humans also gave rise to serious concerns regarding food safety (Bruce *et al.*, 1997).

Breeding with resistant rams

Today classical scrapie is present in all sheep-producing countries except Australia and New Zealand (Detwiler *et al.*, 2003,; Fediaevsky *et al.*, 2008). Infection mostly occurs at very young age and clinical signs of this fatal disease are visible after a variable incubation period of one or more years dependent of genotype (Hadlow *et al.*, 1982). During the incubation period the prion protein PrP^{Sc} slowly accumulate in the animal and can be detected for most sheep genotypes in lymphoid organs such as tonsils before clinical signs become visible (Schreuder *et al.*, 1996, 1998). Scrapie control became a priority in many countries some ten years ago. This was motivated in part by the theoretical possibility that BSE may in past have been introduced into sheep thorough consumption of feed supplements, with potential consequence in public health (Ferguson *et al.*, 2002; Kao *et al*, 2002). This possibility became apparent after experimental infection of sheep with BSE showed that sheep can be infected via the oral route and that the resulting clinical symptoms are very similar to BSE (Foster *et al.*, 2001; Houston *et al*, 2003).

For classical scrapie the most relevant polymorphism for susceptibility occur at codons 136, 154 and 171 of PrP gene (Goldmann *et al.*, 1994; Hunter *et al.*, 1997; Spiropoulos 2007). The VRQ allele confers high susceptibility to most strains of classical scrapie, ARQ allele is associated with moderate susceptibility and AHQ allele may be associated with increased resistance to linger incubation periods (Belt *et al.*, 1995; Bossers *et al.*, 1995). The allele ARR is known to confer resistance to all strains of classical scrapie, with the homozygote genotype ARR/ARR being extremely resistant, and the heterozygote ARQ/ARR and AHQ/ARR genotypes being rarely affected by classical scrapie (Baylis *et al.*, 2004; Hunter *et al.*, 1994). In the Netherland, the selection of resistant rams (ARR/ARR) for breeding was made compulsory for the sheep industry in November 2004.

In six commercially run flocks following this breeding strategy, they used genotyping to monitor the genotype distribution, and tonsil biopsies and post-mortem analysis to monitor the occurrence of scrapie infection. The farmers were not informed about monitoring results until the end of the study period of six years. They used a mathematical model of scrapie transmission to analyse the monitoring data and found that where the breeding scheme was consistently applied, outbreak control was obtained after at most 4 years (Nodelijk *et al.*, 2011) (Table 11.1; C. Bruschke, 2012).

Their results also show that classic scrapie control can be obtained before the frequency of non-resistant animals is reduced zero in the flock. Control at the national scale can be obtained without a loss of genetic polymorphism from any of the sheep breeds.

Results in the Netherlands show that to reach the situation, where R_0 becomes significantly smaller than 1 (as requested for outbreak control) the number of non-ARR/ARR

Table 11.1 Outbreak control of scrapie in Netherlands

	2003	2004	2005	2006	2007	2008	2009	2010
EU[a]								
Slaughtered for consumption	6.6	7.3	8.6	5.6	5.9	7.7	8.9	8.9
Fallen animals	11.3	13.4	23.8	15.7	12.2	14.8	14.6	16.7
Netherlands								
Slaughtered for consumption	21.3	14.5	15.7	6.5	7.6	1.9	1.0	2.0
Fallen animals	15.0	25.6	22.8	14.8	8.7	9.0	2.0	0.0

TSE tests in sheep; positive cases per 10,000 tests.
[a]Excluding Cyprus.

animals does not have to be reduced to zero in the flock (Nodelijk *et al.*, 2011). On a national scale this implies that it will be possible to maintain the diversity of susceptibility alleles desired in case a new scrapie (or BSE) strain targeting ARR would arise in the future.

It is important to note that where a reduction of R_0 to below 1 is achieved, this does not mean that the breeding programme can simply be ceased. Although expected to decline with time, infection risks to susceptible genotype, e.g. from a scrapie-contaminated environment, are expected to remain present for some time. Therefore, re-introducing the use of non-ARR/ARR rams for breeding without risking to revert to a situation $R_0 > 1$ require a new breeding strategy that avoid mating such rams with non-ARR/ARR ewes.

No classical scrapie strain thus far has escaped ARR-associated resistance. In line with this expectation, the breeding programme proved successful in Dutch flocks in 2010 (Melchoir *et al.*, 2010).

As a result this control programme not only the reproduction number R_0, but also the infection pressure (or force of infection) in the field will decrease in time. However, owing to the long incubation period of a scrapie infection, a delay of a few years is expected between the reduction in R_0 and the reduction in infection pressure. This is in line with the detection of new scrapie cases during monitoring study in the Netherlands (Nodelijk *et al.*, 2011), which were all born before the start of the programme. Extending the breeding strategy by a removal of scrapie-susceptible ewes on the basis of their genotype would accelerate the reduction of both R_0 as well as the infection pressure. However, when considering the rapid outbreak control as observed in the Netherlands study, the use of resistant rams seems sufficient and can be recommended as a control strategy in scrapie-affected countries.

Atypical BSE

The question of zoonotic potential of atypical BSE need to be answered. L-type and H-type BSE have so far only occurred in cattle that were at least 8 years old (Biscabe *et al.*, 2004; Casalone *et al.*, 2004). While in L-type BSE, the molecular mass of the lowest protein fragment, representing the unglycosylated form of PrP, migrates slightly lower than that in classical BSE (Casalone *et al.*, 2004), in H-type BSE the same band migrates a little bit higher than in classical BSE (Biacabe *et al.*, 2004). In the initial description of L-type BSE, it had already been reported that the distribution of PrP^{Sc} in the different brain regions did not exactly follow the pattern that was known for classical BSE. This finding may question the possibility to detect such cases, as in weak cases of L-type BSE in the obex and brainstem regions, representing the target areas for all BSE diagnostic methods, may not give positive results in the diagnostic tests. In the brain of the first cow where L-type BSE was detected, the brain stem and obex area did not display the highest PrP^{Sc} concentrations, but instead, abundant amounts were detected in the thalamus and olfactory region (Casalone *et al.*, 2004). Moreover, transmission experiments to Human (Kong *et al.*, 2008) and bovine (Bushmann *et al.*, 2006) transgenic mice as well as macaques (Comoy *et al.*, 2008) had revealed clear indications for a higher zoonotic potential of L-type BSE than classical BSE.

These uncommon cases have been essentially detected in aged asymptomatic cattle during systematic testing at slaughterhouse. In France a retrospective study of all TSE-positive cattle identified through the compulsory EU surveillance programme between 2001–2007 was performed (Biacabe *et al.*, 2008). These studies indicated that:

- All H-BSE and L-BSE cases detected by rapid tests were observed in animals over eight years old in either fallen stock surveillance stream or the abattoir (healthy slaughter).
- No H-BSE and L-BSE were observed in the passive epidemio-surveillance network although, during retrospective interviews, the farmers and veterinarian for six cases of these animals reported clinical signs consistent with TSE in these fallen stock.
- Frequency of H-BSE and L-BSE is respectively 0.35 and 0.41 cases per million adult cattle tested but increases to 1.9 and 1.7 per million in tested animals over 8 years old.

The number of atypical BSE cases detected in countries that have already identified them seems to be comparable from year to year (Table 11.2; F. van Zijderveld, personal communication, 2012; EFSA, 2012).

No comprehensive study on the prevalence of atypical BSE cases has been done in other EU Member States and the performances of the currently available rapid test applied for initial TSE screening of the cattle population towards atypical BSE is still unknown. The true incidence and geographical distribution of atypical forms of BSE has not been reported.

The origin of these atypical BSE cases is currently unknown, as is the performances of the current active surveillance system for detecting H-BSE and L-BSE affected animals, resulting in uncertainty about the real prevalence of these conditions.

All atypical BSE cases identified in the EU were born before the extended or real feed ban that came into a law in January 2001 (Ducrot *et al.*, 2008). Hence, as with classical BSE, exposure of these animals to feed contaminated with low titres of TSE agent cannot be excluded, although other origins for these TSE forms cannot be discarded. In particular, the unusually old ages of all H-BSE and L-BSE identified cases and their apparent low prevalence in the population could suggest that these atypical BSE forms are arising spontaneously.

Table 11.2 Numbers of atypical BSE cases worldwide – 73 cases in total (31 December 2011)

Country	H-type	L-type	Total
Austria	1	2	3
Canada	1	1	2
Denmark	0	1	1
France	13	13	26
Germany	1	1	2
Ireland	3	0	3
Italy	0	5	5
Japan	0	2	2
Poland	2	9	11
Spain	1	0	1
Sweden	1	0	1
Switzerland	1	0	1 and 2 Swiss type
The Netherlands	1	3	4
United Kingdom	4	3	7
USA	2	0	2
Total	31	40	71

To date there is no comprehensive information about the pathogenesis of atypical BSE in cattle. In one Italian study, PrP^{res} was not detected in peripheral nerves (Lombardi *et al.*, 2008), but presence of infectivity in skeletal muscle, presumably linked to nervous structure, of natural cases has been described (Suardi *et al.*, 2009). More recently histopathology showed amyloid deposition in the skeletal muscle of cattle with L-BSE (Suardi *et al.*, 2012). In Japanese study, infectivity and PrP^{res} were detected in many peripheral nerves tested from mid-incubation onwards. The protein was not detected in lymphoid tissues (Iwamaru *et al.*, 2010). There is no result for the distribution of PrP^{res} outside the central nervous system in H-BSE cattle.

Both L-BSE and H-BSE have shown BSE-like characteristics on transmission studies in some line of mice. The precise relationship between classical BSE, H-BSE and L-BSE is not clear. However these experiments have shown that the potential for interspecies transmission of atypical BSE, especially BASE, is high. The transmission barrier observed for the L-BSE was lower than that for classical BSE. In wild-type mice and in transgenic mice expressing the VRQ allele of ovine PrP, the L-BSE agent acquired a phenotype undistinguishable from the BSE agent (Beringue *et al.*, 2007; Capobianco *et al.*, 2007). Transmission of H-BSE isolates originating from France and Poland to bovine-PrP transgenic mice has been reported. While in the majority of the cases the propagated TSE was different from classical BSE, classical BSE has emerged in a proportion of the inoculated mice inoculated with two distinct isolates (one from France and one from Poland) (Espinosa *et al.*, 2010; Baron *et al.*, 2011). In another Spanish study five atypical H-BSE isolates were characterized by analysing their molecular and neuropathological properties during transmission in transgenic mice expressing homologous bovine prion protein (Torres *et al.*, 2011). In several inoculated mice, strain features emerged that were highly similar to those of classical BSE agent. These data indicate that there may be an aetiological relationship between atypical and classical BSE. Besides, These data support the view that the epidemic BSE agent could have originated from atypical BSE prion.

Several elements indicate that the L-BSE agent has potential to be a zoonotic agent. Primates are highly permissive to L-BSE agent, even by oral route, and these can also propagate without any apparent transmission barrier in transgenic mice overexpressing human PrP. Intracerebral inoculation of brain from L-BSE-infected cattle to cynomologus macaque induced a spongiform encephalopathy distinct in all its aspects (clinical, lesional and biochemical) from macaque BSE (Comoy *et al.*, 2008). In the frame of a primary passage through inoculation of a same amount of infected brain, incubation periods were shorter (23–25 months) than for BSE (38–40 months), suggesting that L-BSE may be more virulent than classical BSE for infecting primates. L-BSE was also transmissible to microcebes, with shorter incubation than classical BSE (Baron *et al.*, 2008). Recent experiments demonstrate the transmissibility of L-BSE to macaque by the oral route (Comoy, 2010) with 5 g of infected brain, this amount being similar to the one used for oral transmission of classical BSE in the macaque model. In contrast, no clinical sign has been observed 72 months after intracerebral inoculation of brain from H-BSE, and recipient cynomolgus macaque remained healthy, suggesting a lower, if any, virulence of this agent from primate (Comoy, 2010). Transmission experiments to primates suggest that some TSE agents other than classical BSE agent in cattle (namely L-type atypical BSE, classical BSE in sheep, TME, CWD agents) might have zoonotic potential. In particular, primates are highly permissive to L-type atypical BSE, even by oral route.

In both primates and human PrP transgenic mice models of virulence of the L-BSE agent is significantly higher than that of classical BSE. To date, H-BSE has not been reported as transmissible to mice overexpressing the Met allele of human PrP, nor to primates. The intracerebral inoculation of L-BSE field isolates produced TSE disease in two lines of mice overexpressing human PrP (Met-129), exhibiting a molecular phenotype distinct from classical BSE (Beringue *et al.*, 2008a; Kong *et al.*, 2008). In one of them, the L-BSE agent appeared to propagate with no obvious transmission barrier: a 100% attack rate was observed on the first passage, the incubation time was not reduced on subsequent passaging (Beringue *et al.*, 2008a), and the L-type PrP^{Sc} biochemical signature was essentially conserved (Beringue *et al.*, 2008a, Kong *et al.*, 2008). The latter appeared undistinguishable from that seen after experimental inoculation of MM2 sCJD in these transgenic mice (Beringue *et al.*, 2007). These transmission features markedly differed from the low transmission efficiency of cattle BSE isolates to this (Beringue *et al.*, 2008a,b) and other (Asante *et al.*, 2002) human transgenic mouse lines.

H-type isolates failed to infect one line of 'humanized' mice (Beringue *et al.*, 2008a). These mice overexpress human PrP and were inoculated intracranially with a low dilution inoculums, supporting the view that the transmission barrier of H-type BSE from cattle to humans might be quiet robust. The permissiveness of 'humanized' transgenic mice expressing the valine allele at codon 129 (or hemizygous) to atypical BSEs is currently unknown.

There is uncertainty on the origin of sporadic CJD. There is speculation that sCJD is due to exposure to atypical BSE, or is linked to classical scrapie, atypical scrapie etc. This uncertainty is the subject of risk assessment of food safety (EFSA Panel on Biological Hazards (BIOHAZ), 2011). Besides there is uncertainty on 'origin of BSE'. The type and origin of the ('scrapie-like', L-BSE or H-BSE) prions that initially caused cases of BSE in cattle are unknown, and unlikely to be elucidated, but the identification of MBM (and possibly other by-products, tallow etc.) in animal feed as the 'extended' common source of the epidemic is an epidemiological fact [EFSA Panel on Biological Hazards (BIOHAZ), 2011].

PMCA and prion-associated activities

It will be intellectually more satisfying to know exactly what the infectious agent at the level of molecular architecture and accessory molecules (lipids, nucleic acids and carbohydrate). Even without this knowledge, it can be accepted, based on the results from *de novo* generation and other *in vitro* conversion experiments, and transmission experiments to human PrP transgenic mice, that the human prion protein can be converted to a PrP^{Sc}-like form by animal PrP^{Sc}: there is not absolute barrier to infection/conversion of humans by mammalian TSEs at the molecular level. Irrespective of the precise nature of the infectious agent, there are sufficient data to say that animal prions have a potential to infect humans. Experimentally, prion-associated seeding activity converting normal protease- sensitive PrP to protease K-resistant prion protein (PrP^{res}) can be monitored *in vitro* by protein misfolding cyclic amplification (PMCA) (Saborio *et al.*, 2001; Castilla *et al.*, 2006). Serial PMCA (Bieschke *et al.*, 2004; Saa *et al.*, 2005) has been established during the past few years as powerful tool for the ultrasensitive – yet generally non-quantitative – detection of minute amount of PrP^{TSE}. Chen *et al.* (2010) and Wilham *et al.* (2010) described two technical advancement, called quantitative PMCA (qPMCA) and real-time quaking induced conversion assay (RT-QuIC), which showed that the estimation of prion titres and prion seeding activity, respectively, are biochemically feasible *in vitro* with high sensitivity and accuracy. Prion infectivity can be titrated also biologically *in vitro*, as was exemplified by cell culture assay for the quantitative detection of RML prions in solutions or on steel wire used as model carriers for disinfection (Kolohn *et al.*, 2003; Mahal *et al.*, 2008; Edgeworth *et al.*, 2009).

Prions have a particularly high tolerance to inactivation (Taylor, 2000; Taylor, 2004). Therefore, they constitute a complex challenge to the safe maintenance of re-usable surgical instruments and medical devices (Beekes *et al.*, 2004). However, as recently shown, this challenge can be turned into benefit when prions are exploited as an informative paradigm for the development of novel disinfectants that are simultaneously active against bacteria, viruses as well as fungi (Lehmann *et al.*, 2009, Beekes *et al.*, 2010). In order to facilitate the use of prions as model agents in the search for novel broad-range disinfectants Pritzkow *et al.* established an experimental platform for the sensitive measurement as biological detection of scrapie seeding activity *in vitro* (Pritzkow *et al.*, 2011).

In the study of Pritzkow *et al.* (2011), scrapie seeding activity on prion-contaminated steel wire processed for decontamination was quantified by specially adapted PMCA. PMCA has been used previously by other groups as a rapid test for the assessment of prion inactivation (Murayama *et al.*, 2006; Suyama *et al.*, 2007), however, not in a quantitative way. PMCA needs to be highly standardized and robust in terms of a consistent and objectively quantifiable PrP^{res} amplification if to be used for a quantification of the proteinaceous seeding activity of prions. Recently it was reported that conducting PMCA in the presence of Teflon beads significantly improves the yield, rate and robustness of PrP conversion seeded by 263 K scrapie prions (Gonzales-Montalban *et al.*, 2011). They reported that the diameter of Teflon beads was critical importance for the bead effect of PMCA.

Consistent with the quantitative correlation between scrapie seeding activity and observed in Pritzkow's study (2011), Wilham *et al.* (2010) reported results that also pointed to a direct quantitative correspondence between the seeding and infectious activities of 263K scrapie prions measures by RT-QuIC and bioassay. This provides a remarkable concurrence given the methodological difference of the two studies, and seems further substantiate the association between these prion-associated activities. Additional studies are necessary to further substantiate the observed association between the seeding activity and biological infectivity of 263K prions for other disinfectant formulations and modes of inactivation. In addition, the methodological, conceptual and practical results described in the report of 263K scrapie prions should be validated for the most relevant human TSE agents (Peretz *et al.*, 2006; Giles *et al.*, 2008); such as prions associated with sporadic CJD [sCJD]/subtypes MM1 or VV2 (Heinemann *et al.*, 2007), or variant CJD) on different types of carriers or surfaces (Lipscomb *et al.*, 2006). For this purpose, PMCA protocols and cell culture protocols need to be adapted for the propagation and detection of PrP seeding activities associated with sCJD or vCJD prions.

References

Asante, E.A., Linehan, J.M., Desbruslais, M., Joiner, S., Gowland, I., Wood, A.L., Welch, J., Hill, A.F., Lloyd, S.E., Wadsworth, J.D.F., *et al.* (2002). BSE propagate as either variant CJD-like or sporadic CLD-like prion strains in transgenic mice expressing human prion protein. EMBO J. *21*, 6358–6366.

Baron, T. (2011). Emergence of classical BSE strain properties during serial passages of H-BSE in wild type mice. Plos One *6*, e15389.

Baron, T., Biacabe, A.G., Rouland, S., Verdier, J.M., and Mestre-Francis, N. (2008). Transmission of atypical BSE to Microcebus murinus, a non-human primate: development of clinical symptoms and tissue distribution of PrPres. Proceeding Prion 2008 Conference, Madrid. 8–10 October 2008, p. 16.

Beekes, M., Mielke, M., Pauli, G., Baier, M., and Kurth, R. (2004). Aspect of risk assessment and risk management of nosocomial transmission of classical and variant Creutzfeldt–Jakob disease with special attention to German regulations. Contrib. Microbiol. *11,* 117–135.

Beekes, M., Lemmer, K., Thomizig, A., Jonic, M., Tintelnot, K., and Mielke, M. (2010). Fast, broad-range disinfection of bacteria, fungi, viruses and prions. J. Gen. Virol. *91,* 580–589.

Beringue, V., Andreoletti, O., Le Dur, A., Essalmani, R., Vilotte, J.L., Lacroux, C., Reine, F., Herzog, L., Biacabe, A.G., Caramrlli, M., *et al.* (2007). A bovine prion aquires an epidemic bovine spongiform encephalopathy strain-like phenotype on interspecies transmission. J. Neurosci. *27,* 6965–6971.

Beringue, V., Herzog, L., Reine, F., Le Dur, A., Casalone, C., Vilotte, J.L., and Laude, H. (2008a) Transmission of atypical bovine prions to mice transgenic for human prion protein. Emerg. Infect. Dis. *14,* 1898–1901.

Beringue, V., Le Dur, A., TixaDor, P., Reine, F., Leourry, L., Perret-Liaudet, A., Haik, S., Violtte, J.L., Fontes, M., and Laude, H. (2008b). Prominent and persistent extraneural infection in human PrP transgenic mice infected with variant CJD. PloS One *3,* 1.

Biacabe, A.G., Laplanche, J.L., Ryder, S., and Baron, T. (2004). Distinct molecular phenotypes in bovine prion diseases. EMBO Rep. *5,* 110–115.

Biacabe, A.G., Morignat, E., Win, J., Calavas, D., and Baron, T.G.M. (2008). Atypical bovine spongiform encephalopathies, France, 2001–2007. Emerg. Infect. Dis. *14,* 298–300.

Bieschke, J., Weber, P., Sarafoff, N., Beekes, M., Giese, A., and Kretzschmar, H. (2004). Autocatalytic self-propagation of misfolded prion protein. Proc. Natl. Acad. Sci. U.S.A. *101,* 12207–12211.

Bruce, M.E., Will, R.G., Ironside, J.W.W., McConnell, I., Drummond, D., Suttie, A., McCardle, L., Chree, A., Hope, J., Brickett, C.R., *et al.* (1997). Transmission to mice indicate that 'new variant' CJD is caused by the BSE agent. Nature *389,* 498–501.

Bruschke, C. (2012). Opening speech Prion 2012 conference, Amsterdam, 10 May 2012.

Buschmann, A., Gretzschel, A., Biacabe, A.G., Corona, C., Hoffmann, C., Eiden, M., Baron, T., Caramelli, M., Conraths, F.J., and Groschup, M.H. (2006). Atypical BSE cases in Germany. Vet. Microbiol. *117,* 103–116.

Capobianco, R., Casalone, C., Suardi, S., Mangieri, M., Miccolo, C., Limido, L., Catania, M., Rossi, G., Di Fede, G., Giaccone, G., *et al.* (2007). Conversion of BASE prion strain into the BSE strain: The origin of BSE? PloS Pathog. *3,* e31.

Casalone, C., Zanusso, G., Acutis, P., Ferrari, S., Capucci, L., Tagliavini, F., Monaco, S., and Caramelli, M. (2004). Identification of a second bovine amyloidotic spongiform encephalopathy: molecular similarities with sporadic Creutzfeldt–Jakob disease. Proc. Natl. Acad. Sci. U.S.A. *101,* 3065–3070.

Castilla, J., Saa, P., Morales, R., Abid, K., Maundrell, K., and Soto, C. (2006). Protein misfolding cyclic amplification for diagnosis and prion propagation studies. Methods Enzymol. *412,* 3–21.

Chen, B., Morales, R., Barria, M.A., and Soto, C. (2010). Estimating prion concentration in fluids and tissues by quantitative PMCA. Nat. Methods *7,* 519–520.

Comoy, E. (2010). Transmission studies in primates. Workshop on the epidemiology of human and animal TSEs. 30 April 2010, Torino, Italy.

Comoy, E.E., Casalone, C., Lescoutra-Etchegaray, N., Zanusso, G., Freire, S., Marce, D., Auvre, F., Ruchoux, M.M., Ferrari, S., Monaco, S., *et al.* (2008). Atypical BSE (BASE) transmitted from asymptomatic aging cattle to a primate. PLos One 3, e3017.

Detweiler, L.A., and Baylis, M. (2003). The epidemiology of scrapie. Rev. Sci. Tech. *22,* 121–143.

Ducrot, C., Arnold, M., de Koeijer, A., Heim, D., and Calavas, D. (2008). Review on the epidemiology and dynamics of BSE epidemics. Vet. Res. *39,* 4.

Edgeworth, J.A., Jackson, G.S., Clarke, A.R., Weismann, C., and Collinge, J. (2009). Highly sensitive quantitative cell-based assay for prions absorbed to solidify surface. Proc. Natl. Acad. Sci. U.S.A. *106,* 3479–3483.

European Food Safety Authority (2012). Scientific and technical assistance on the minimum sample size to test should an annual BSE statistical testing regime be authorized in healthy slaughtered cattle. EFSA J. *10,* 2913–3013.

EFSA Panel on Biological Hazards (BIOHAZ). (2011). Joint Scientific Opinion on any possible epidemiological or molecular association between TSEs in animals and humans. EFSA J. *9,* 1945–2055.

Espinosa, J.C., Andreoletti, O., Lacroux, C., Prieto, I., Lorenzo, P., Larska, M., Baron, T., and Torres, J.M. (2010). Atypical H-type BSE infection in bovine-PrP transgenic mice let to the emergence of classical BSE strain features. Prion *4,* 137.

Fediaevsky, A., Tongue, S.C., Noremark, M., Calavas, D., Ru, G., and Hopp, P. (2008). A descriptive study of the prevalence of atypical and classical scrapie in sheep in 20 European countries. BMC. Vet. Res. *4,* 19.

Ferguson, N.M., Ghani, A.C., Donnelly, C.A., Hagenaars, T.J., and Anderson, R.M. (2002). Estimating human health risk form possible BSE infection of the British sheep flock. Nature *415*, 420–424.

Foster, J.D., Parnham, D., Chong, A., Goldmann, W., and Hunter, N. (2001). Clinical signs, histopathology and genetics of experimental transmission of BSE and natural scrapie to sheep and goat. Vet. Rec. *148*, 165–171.

Giles, K., Glidden, D.V., Beckwith, R., Seoanes, R., Peretz, D., DeArmond, S.J., and Prusiner, S.B. (2008). Resistance of bovine spongiform encephalopathy (BSE) prions to inactivation. Plos Pathog. *4*, e1000206.

Goldmann, W., Hunter, N., Smith, G., Foster, J., and Hope, J. (1994). Prp genotype and agent effects in scrapie: change in allelic interaction with different isolates of agent in sheep, a natural host of scrapie. J. Gen. Virol. *75*, 989–995.

Gonzales-Montalban, N., Makarava, N., Ostapchenko, V.G., Savtchenk, R., Alexeeva, I., Rower, R.G., and Baskakov, I.V. (2011). Highly efficient protein misfolding cyclic amplification. Plos Pathog. *7*, e1001277.

Hadlow, W.J., Kennedy, R.C., and Race, R.E. (1982). Natural infection of suffolk sheep with scrapie virus. J. Infect. Dis. *146*, 657–664.

Houston, E.F., and Gravenor, M.B. (2003). Clinical signs in sheep experimentally infected with scrapie and BSE. Vet. Rec. *152*, 333–334.

Hunter, N. (1997). PrP genetics in sheep and the implications for scrapie and BSE. Trends. Microbiol. *5*, 331–334.

Iwamaru, Y., Imamura, M., Matsuura, Y., Masujin, K., Shimizu, Y., Shu, Y.J., Kurachi, M., Kasai, K., Murayama, Y., Fukuda, S., *et al.* (2010). Accumulation of L-type bovine prions in peripheral nerve tissue. Emerg. Infect. Dis. *16*, 1151–1154.

Kao, R.R., Gravenor, M.B., Baylis, M., Bostock, C.J., Chitota, C.M., Evans, J.C., Goldmann, W., Smith, A.J.A., and McLean, A.R. (2002). The potential size and duration of an epidemic of bovine spongiform encephalopathy in British sheep. Science *295*, 332–335.

Klohn, P.C., Stoltze, L., Flechsig, E., Enari, M., and Weissmann, C. (2003). A quantitative, highly sensitive cell-based infectivity assay for mouse scrapie prions. Proc. Natl. Acad. Sci. U.S.A. *100*, 11666–11671.

Kong, Q., Zheng, M., Canalone, C., Qing, L., Huang, S., Chakraborty, B., Wang, P., Chen, F., Cali, I., Corona, C., *et al.* (2008). Evaluation of the human transmission risk of a atypical bovine spongiform encephalopathy prion strain. J. Virol. *82*, 3697–3701.

Mahal, S.P., Demczyk, C.A., Smith, E.W., Klohn, P.C., and Weissmann, C. (2008). Assaying prions in cell culture: the standard scrapie cell assay (SSCA) and the scrapie cell assay in end point format (SCEPA). Methods Mol. Biol. *459*, 49–68.

Melchoir, M.B., Windig, J.J., Hagenaas, T.J., Bossers, A., Davidse, A., and Van Zijderveld, F.G. (2010). Eradication of scrapie with selective breeding: are we nearly there? Vet. Res. *6*, 24–33.

Murayama, Y., Yoshioka, M., Horii, H., Takata, M., Yokoyama, T., Sudo, T., Sato, K., Shinagawa, M., and Mohri, S. (2006). Protein misfolding cyclic amplification as a rapid test for assessment of prion inactivation. Biochem. Biophys. Res. Comm. *348*, 758–762.

Lehmann, S., Pastore, M., Rogez-Kreuz, G., Richard, M., Belondrade, M., Rauwel, G., Durand, F., Yousfi, R., Criquelion, J., Clayette, P., *et al.* (2009). New hospital disinfection process for both conventional and prion infectious agents compatible with thermosensitive equipment. J. Hosp. Infect. 72, 342–350.

Lombardi, G., Casalone, C., D'Angelo, A., Delmetti, D., Torcoli, G., Barbieri, I., Corona, C., Fasoli, E., Farinazzo, A., Fiorini, M., *et al.* (2008). Intraspecies transmission of BASE induces clinical dullness and amyotrophic changes. Plos Pathog. *4*, 5.

Nodelijk, G., van Roermund, H.J.W., van Keulen, L.J.M., Engel, B., Vellema, P., and Hagenaars, T.J. (2011). Breeding with resistant rams leads to rapid control of classical scrapie in affected flocks. Vet. Res. *42*, 5–16.

Peretz, D., Supattapone, S., Giles, K., Vergara, J., Freyman, Y., Lessard, P., Safar, J.G., Glidden, D.V., McCulloch, C., Nguyen, H.O., *et al.* (2006). Inactivation of prions by acidic sodium dodecyl sulfate. J. Virol. *80*, 322–331.

Pritzkow, S., Wagenfuhr, K., Daus, M.L., Boerner, S., Lemmer, K., Thomzig, A., Mielke, M., and Beekes, M. (2011). Quantitative detection and biological propagation of scrapie seeding activity *in vitro* facilitate use of prions as model pathogens for disinfection. Plos One *6*, e20384.

Saa, P., Castilla, J., and Soto, C. (2005). Cyclic amplification of protein misfolding and aggregation. Methods Mol. Biol. *299*, 53–65.

Saborino, G.P., Permanne, B., and Soto, C. (2001). Sensitive detection of pathological prion protein by cyclic amplification of protein misfolding. Nature *411*, 810–813.

Schreuder, B.E., van Keulen, L.J., Vromans, M.E., Langeveld, J.P., and Smits, M.A. (1996). Preclinical test for prion diseases. Nature *381*, 563.

Schreuder, B.E., van Keulen, L.J., Vromans, M.E., Langeveld, J.P., and Smits, M.A. (1998). Tonsillar biopsy and PrPSc detection in the preclinical diagnosis of scrapie. Vet. Rec. *142*, 564–568.

Suardi, S., Vimercati, C., Moda, F., Ruggerone, M., Campagnani, I., Lombardi, G., Gelmetti, D., Groschup, M., Buschmann, A., Casalone, C., *et al.* (2009). Infectivity in skeltal muscle of BASE-infected cattle. Proceeding Prion 2009 Conference. Thessaloniki, 23–25 September 2009, p. 47.

Suardi, S., Vimercati, C., Casalone, C., Gelmetti, D., Corona, C., Iulini, B., Mazza, M., Lombardi, G., Moda, F., Ruggerone, M., *et al.* (2012). Infectivity in skeltal muscle of cattle with atypical bovine spongiform encephalopathy. Plos One 7, e31449.

Suyama, K., Yoshioka, M., Akagawa, M., Murayama, Y., Horii, H., Tanaka, M., Yokoyama, T., and Mohri, S. (2007). Prion inactivation by the Maillard reaction. Biochem. Biophys. Res. Communn. *356*, 758–762.

Taylor, D.M. (2000). Inactivation of transmissible degenerative encephalopathy agents: A review. Vet. J. *159*, 10–17.

Taylor, D.M. (2004). Resistance of transmissible spongiform encephalopathy agents to decontamination. Contrib. Microbiol. *11*, 136–145.

Torres, J.M., Andreoletti, O., Lacroux, C., Prieto, I., Lorenzo, P., Larska, M., Baron, T., and Espinosa, J.C. (2011). Classical bovine spongiform encephalopathy by transmission of H-type prion in homologus prion protein context. Emerg. Infect. Dis. *17*, 1636–1644.

Wells, H.A.H., Scott, A.C., Johnson, C.T., Gunning, R.F., Hanock, R.D., Jeffrey, M., Bradley, R., and Dawson, M. (1987). A novel progressive spongiform encephalopathy in cattle. Vet. Rec. *121*, 419–420.

Wilham, J.M., Orru, C.D., Bessen, R.A., Atarashi, R., Sano, K., Race, B., Meade-White, K.D., Taubner, L.M., Timmes, A., and Caughey, B. (2010). Rapid end-point quantitation of prion seeding activity with sensitivity comparable to bioassays. Plos Pathog. *6*, e1001217.

Index

A

B

C

D

E

F

G

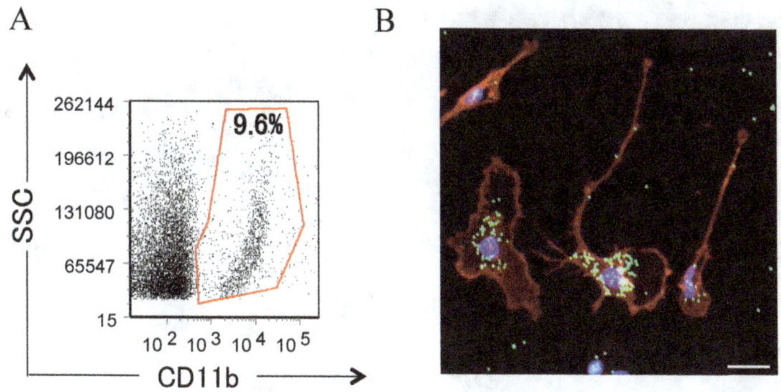

Figure 4.1 Population and phagocytosis of microglia. (A) The rate of microglia (CD11b-positive cells) in the brain cells is approximate 10%. (B) Microglia showed phagocytic activity and activated microglia released inflammatory cytokines. Green shows latex beads, red shows CD11b, and blue shows nuclei. Scale bar is 20μm.

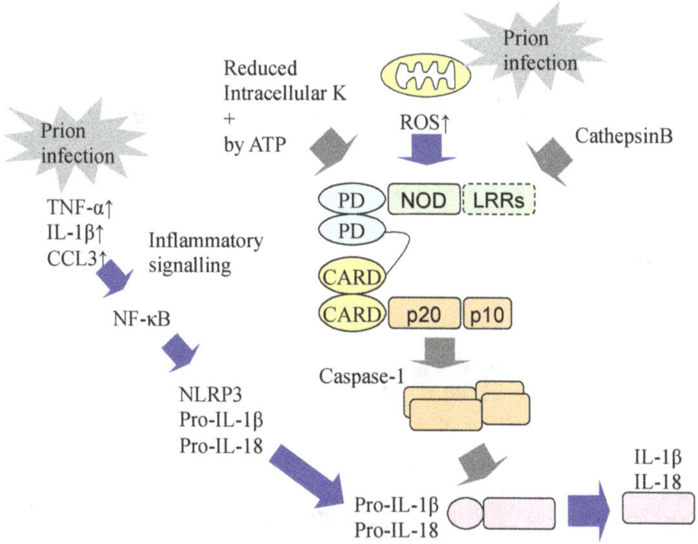

Figure 4.2 The schematic pathway of NLRP3 Inflammasome activation. IL-1β is produced via NLRP3 inflammasome activation. NLRP activation is associated with ROS production in mitochondria.

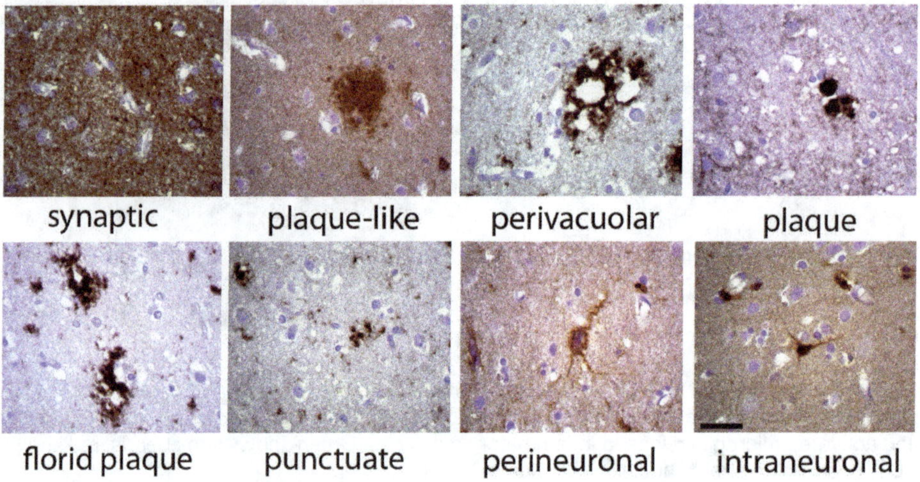

Figure 5.1 Deposition patterns of abnormally folded prion protein in the brain. Shown are representative examples of deposition patterns described for the deposition of abnormally folded prion protein. These only become visible upon immunolabelling with antibodies directed against the prion protein. Nomenclature of deposition patterns is indicated below histology. Scale bar represents 50μm.

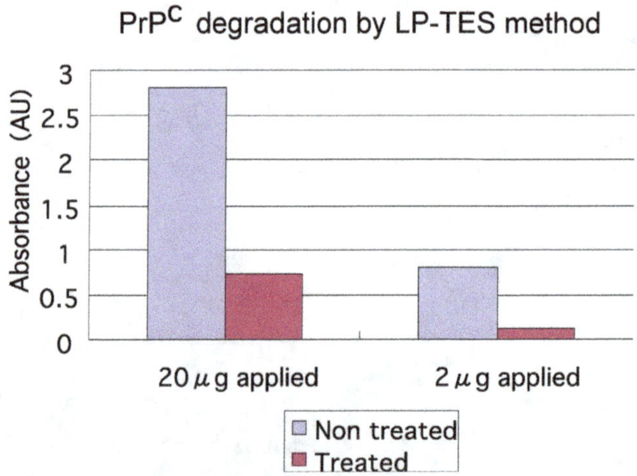

Figure 6.2 20μg or 2μg of PrP^c was applied. After nitrogen gas plasma exposure, the residue PrP^c was sampled with EIA buffer and assayed with an ELISA kit from Funakoshi, Tokyo, Japan. Nitrogen gas plasma exposure condition: (low pressure method) (reactor); anode–cathode gap: 40mm; vacuum: 43,000Pa; nitrogen gas flow speed: 6l/min; time: 30 minutes; temperature: 60°C (temperature was measured inside of plasma); (electric source) input power: 84W; rep. freq. 2.5kHz; peak voltage: 19.0kV. The smaller the PrP^c, the more degradation is observed. This indicates clump phenomenon of PrP^c onto carrier materials can be formed. LP-TES, low-pressure triple effects sterilization; ELISA, enzyme-linked immunosorbent assay; EIA, enzyme immunoassay.